Differential Forms Orthogonal to Holomorphic Functions or Forms, and their Properties

TRANSLATIONS OF MATHEMATICAL MONOGRAPHS

VOLUME **56**

Differential Forms Orthogonal to Holomorphic Functions or Forms, and their Properties

by L. A. Aĭzenberg
Sh. A. Dautov

American Mathematical Society · Providence · Rhode Island

ДИФФЕРЕНЦИАЛЬНЫЕ ФОРМЫ,
ОРТОГОНАЛЬНЫЕ ГОЛОМОРФНЫМ
ФУНКЦИЯМ ИЛИ ФОРМАМ,
И ИХ СВОЙСТВА

Л. А. АЙЗЕНБЕРГ И Ш. А. ДАУТОВ

ИЗДАТЕЛЬСТВО «НАУКА»
СИБИРСКОЕ ОТДЕЛЕНИЕ
НОВОСИБИРСК 1975

Translated from the Russian by R. R. Simha
Translation edited by Lev J. Leifman

2000 *Mathematics Subject Classification.* Primary 32–XX.

ABSTRACT. This book is devoted to the description of exterior differential forms orthogonal to holomorphic forms of degree $n-p, 0 \le p \le n$ (in particular, to holomorphic functions if $p=n$) with respect to integration over the boundary of a bounded domain D in $\mathbf{C}^n$. The Martinelli-Bochner-Koppelman formula, which is an integral representation of exterior differential forms, is given, and the characteristic properties of the trace of a holomorphic function on the boundary are studied. The question of representation and multiplication of distributions lying in $\mathcal{D}'(\mathbf{R}^{2n-1})$ is discussed with the aid of $\bar\partial$-closed forms of type $(n, n-1)$ with harmonic coefficients.

Library of Congress Cataloging-in-Publication Data

Aĭzenberg, Lev Abramovich, 1937–
Differential forms orthogonal to holomorphic functions or forms, and their properties.
(Translations of mathematical monographs; v. 56)
Translation of: Different͡sial′nye formy, ortogonal′nye golomorfnym funkt͡sii͡am ili formam, i ikh svoĭstva.
Bibliography: p.
Includes index.
1. Holomorphic functions. 2. Exterior forms. 3. Differential forms. I. Dautov, Sh. A. (Shamil′ Abdullovich) II. Title. III. Series.

QA331.A46313 1983 515.9 83-2696
ISBN 0-8218-4508-X
ISSN 0065-9282

AMS softcover ISBN 978-0-8218-1348-5

10 9 8 7 6 5 4 3 2 05 04 03 02 01 00

TABLE OF CONTENTS

PREFACE TO THE AMERICAN EDITION

In the six years that have elapsed since this book appeared in the USSR, many new results have been obtained in this field of multidimensional complex analysis. These results are presented in a supplement (Chapter V–VII), written by the authors especially for the American edition. The results of A. M. Kytmanov have made the greatest impact on the contents of the supplement. He has also written Chapter VII of the supplement at the request of the authors. We take this opportunity to thank him for this work; we also thank him and S. G. Myslivech for help in preparing the manuscript of the supplement.

PREFACE

In this book we consider the problem of characterizing the exterior differential forms which are orthogonal to holomorphic functions (or forms) in a domain $D \subset \mathbf{C}^n$ with respect to integration over the boundary, and some related questions. We give a detailed account of the derivation of the Bochner-Martinelli-Koppelman integral representation of exterior differential forms, which was obtained recently (1967) but has already found many important applications. A complete proof of this representation has not previously been available in our* literature. We study the properties of $\bar{\partial}$-closed forms of type $(p, n-1)$, $0 \leqslant p \leqslant n-1$, which turn out to be the duals (with respect to the orthogonality mentioned above) to holomorphic functions (or forms) in several complex variables, and resemble holomorphic functions of one complex variable in their properties. At the end of the book, we give some applications, in particular to the problem of multiplying distributions, and also a brief historical survey and a discussion of open problems.

We hope that this little book will be useful to mathematicians and theoretical physicists interested in several complex variables.

The greater part of the results expounded below were obtained by us during the years 1970–73. They were reported in seminars at Krasnoyarsk, Moscow State University, Urals State University (Sverdlovsk), and the Institute of Mathematics in the Siberian Division of the Academy of Sciences of the USSR. We thank the participants of these seminars for useful discussions. Above all we are grateful to G. M. Khenkin and V. P. Palamodov for valuable remarks.

* *Editor's note.* The authors mean "Russian".

INTRODUCTION

A large number of results in the theory of holomorphic functions of one complex variable depend on the following, already classical, assertion:

Let D be a bounded domain in the plane with smooth boundary ∂D, and $\varphi(z)$ a continuous function on ∂D. Then

$$\int_{\partial D} f(z)\varphi(z)\,dz = 0 \tag{0.1}$$

for all functions $f(z)$ continuous in the closed domain $\overline{D}$ and holomorphic in D if and only if $\varphi(z)$ can be extended holomorphically into D (see, for example, Mushelishvili [1], §29). Briefly, holomorphic functions, and only these, are orthogonal to holomorphic functions.

By attaching the dz in (0.1) to $f(z)$ or to $\varphi(z)$, the same result can be reformulated in the following way, which is more convenient for our exposition: *a continuous (on ∂D) exterior differential form*(1) *α of type $(p,0)$ is orthogonal under integration on ∂D to all forms of type $(1-p,0)$ which are continuous in $\overline{D}$ and holomorphic in D, if and only if there exists a form γ of type $(p,0)$ holomorphic in D and continuous in $\overline{D}$, such that $\alpha = \gamma\,|_{\partial D}$. Here $p = 0$ or 1.*

The analogous problem is of great interest in the theory of functions of several complex variables:

What forms α, defined on the boundary of a domain D in $\mathbf{C}^n$ with smooth boundary, are orthogonal to holomorphic forms μ (in particular to holomorphic functions) in the sense that

$$\int_{\partial D} \alpha \wedge \mu = 0 \tag{0.2}$$

for all forms holomorphic on $\overline{D}$ (or holomorphic in D and continuous on $\overline{D}$)?

---

(1) In the sequel, we shall simply write "form" instead of "exterior differential form".

Kohn and Rossi [1] have considered the more general problem of describing forms which are orthogonal to all $\bar\partial$-closed forms of type $(n-p, q)$. They have shown that, if $q > 0$, and if at each point of ∂D the Levi form has $q+1$ positive eigenvalues or $n-q$ negative ones, then a C^∞ form α of type $(p, n-q-1)$ on ∂D is orthogonal (in the sense of (0.2)) to all $\bar\partial$-closed C^∞ forms of type $(n-p, q)$ on D if and only if there exists a C^∞ $\bar\partial$-closed form γ of type $(p, n-q-1)$ on $\bar D$ such that $\gamma|_{\partial D} = \alpha$.

However this solution does not include the case of interest to us, since holomorphic forms are $\bar\partial$-closed forms of type $(p, 0)$.

It will be proved (Chapter II) that $\bar\partial$-closed forms of type $(p, n-1)$, $0 \leq p \leq n$, are the duals, with respect to integration over the boundary of the domain (in the sense of (0.2)), of holomorphic forms of type $(n-p, 0)$; in particular, for $p = n$, they are the duals of holomorphic functions.

The proof makes essential use of the Martinelli-Bochner-Koppelman integral representation of exterior differential forms, as also the "barrier function" of Khenkin for strictly pseudoconvex domains.

To state the result in a concrete way: forms which are orthogonal to all holomorphic forms on the closure of a strictly pseudoconvex domain must be extendable to the interior as $\bar\partial$-closed forms, or (what amounts to the same in the case of strictly pseudoconvex domains) as $\bar\partial$-exact forms of type $(p, n-1)$.

On the other hand, it will be established in Chapter III that the $\bar\partial$-closed forms of type $(p, n-1)$, $0 \leq p \leq n$, are the analogs for $\mathbf{C}^n$ of the holomorphic functions of one complex variable, i.e. they retain several of the important properties of these functions which e.g. the holomorphic functions of n variables, $n > 1$, do not possess.

Chapter I is devoted to tools that will be used and results related to the main problem. In particular, we present an elementary derivation of the Martinelli-Bochner-Koppelman integral representation of forms, generalizing the Martinelli-Bochner integral representation for holomorphic functions on the one hand, and the Cauchy-Green formula for smooth functions in the case $n = 1$ on the other. Also we give a characterization of the trace of a holomorphic function on the boundary of a domain in $\mathbf{C}^n$.

In Chapter IV, some applications are indicated: we generalize the classical theorems of Hartogs and the Rieszes, describe the general form of integral representations of holomorphic functions, and construct the Martinelli-Bochner representation of a distribution lying on $\mathfrak{D}'(\mathbf{R}^{2n-1})$, by means of which we introduce a new definition for the product of distributions.

In the text itself, we shall refer only to such original works as are necessary for understanding the book. All references are listed at the end of the book. After Chapters IV and VII the reader will also find a brief historical survey; in the former some unresolved questions (as of 1975) are indicated (a few of which are subsequently solved in Chapters V–VII, written in 1981 for the American edition).

We shall employ the following notation. $\mathbf{C}^n$ is the space of n complex variables, whose points are denoted by z, ζ, z^0, ζ^0, etc. If $z = (z_1,\ldots,z_n)$ then $\bar{z} = (\bar{z}_1,\ldots,\bar{z}_n)$. For $z, \zeta \in \mathbf{C}^n$ we define $\langle z, \zeta\rangle = z_1\zeta_1 + \cdots + z_n\zeta_n$, and

$$|z| = \sqrt{\langle z, \bar{z}\rangle}, \qquad t(\zeta, z) = \frac{\bar{\zeta} - \bar{z}}{|\zeta - z|^2} = \left(\frac{\bar{\zeta}_1 - \bar{z}_1}{|\zeta - z|^2},\ldots,\frac{\bar{\zeta}_n - \bar{z}_n}{|\zeta - z|^2}\right),$$

$$B(z^0, r) = \{z \in \mathbf{C}^n : |z - z^0| < r\}.$$

For vectors I and J with nonnegative integral components,

$$\|I\| = i_1 + \cdots + i_n, \qquad z^I = z_1^{i_1} \cdot \cdots \cdot z_n^{i_n},$$

$$D^{I,J} = \frac{\partial^{\|I-J\|}}{\partial z_1^{i_1} \cdots \partial z_n^{i_n} \partial \bar{z}_1^{j_1} \cdots \partial \bar{z}_n^{j_n}}.$$

Σ'_J denotes summation over $J = (j_1,\ldots,j_p)$ with $1 \leqslant j_1 < \cdots < j_p \leqslant n$. Also,

$$dz = dz_1 \wedge \cdots \wedge dz_n, \qquad dz_J = dz_{j_1} \wedge \cdots \wedge dz_{j_p},$$

$$dz[J] = dz_{i_1} \wedge \cdots \wedge dz_{i_{n-p}}, \qquad i_1 < \cdots < i_{n-p},$$

$$j_k \neq i_l, \qquad k = 1,\ldots,p, \qquad l = 1,\ldots,n-p.$$

The symbol [ ] will often be employed in the above sense; e.g. $z[k] = (z_1,\ldots,z_{k-1}, z_{k+1},\ldots,z_n)$.

Let M and N be sets in $\mathbf{C}^n$. Then ∂M is the boundary of M, and we define

$$\rho(z, M) = \inf_{\zeta \in M} |\zeta - z|, \qquad M^r = \{z \in M : \rho(z, \partial M) > r\},$$

$$\operatorname{diam} M = \sup_{\zeta, z \in M} |\zeta - z|.$$

CM denotes the complement of M in $\mathbf{C}^n$.

$\overline{M}$ is the closure of M, int M is the interior of M, and $M \Subset N$ means that $\overline{M}$ is compact and contained in int N.

Let F be a closed set and Ω an open set in $\mathbf{C}^n$. For $m = 0, 1,\ldots,\infty$, $0 \leqslant \lambda \leqslant 1$, $0 \leqslant r \leqslant 2n$, and $0 \leqslant p, q \leqslant n$, $C_r^{m,\lambda}(F)$ denotes the class of forms of dimension r whose coefficients extend continuously to C^m functions in a neighborhood of F such that all derivatives of order m (if $m < \infty$) satisfy a Hölder condition with exponent λ.

$C_{(p,q)}^{m,\lambda}(F)$ denotes the subset of $C_{p+q}^{m,\lambda}(F)$ consisting of forms of type (p, q).

For $\lambda = 0$ we shall omit the index, and regard $C_{(p,q)}^m(F) = C_{(p,q)}^{m,0}(F)$ as a space with the topology of uniform convergence of all coefficients together with all derivatives of order $\leqslant m$ on F.

$C^{m,\lambda}_{(p,q)}(\Omega)$ denotes the projective limit of the classes $C^{m,\lambda}_{(p,q)}(F)$, $F \Subset \Omega$, with the projective limit topology.

If

$$\alpha = \sum_{I,J}{}' a_{IJ} dz_I \wedge d\bar{z}_J \in C^{m,\lambda}_{(p,q)}(\Omega)$$

then

$$\bar{\partial}\alpha = \sum_{I,J}{}' \sum_{k=1}^{n} \frac{\partial a_{IJ}}{\partial \bar{z}_k} d\bar{z}_k \wedge dz_I \wedge dz_J,$$

$$Z^{m,\lambda}_{(p,q)}(\Omega) = \left\{\alpha \in C^{m,\lambda}_{(p,q)}(\Omega) : \bar{\partial}\alpha = 0\right\}$$

(for $m = 0$, differentiation is to be understood as that in the space of distributions).

Here again, if $m = 0$ we shall omit the index, and write e.g. $Z^0_{(p,q)}(\Omega) \equiv Z_{(p,q)}(\Omega)$. Similarly we shall drop the indicies p and q if $p = q = 0$.

For all $\alpha \in C_{(p,q)}(\Omega)$ we set $\operatorname{supp} \alpha = \overline{\{z \in \Omega : \alpha(z) \neq 0\}}$,(²) and

$$A_p(\Omega) = Z_{(p,0)}(\Omega) = Z^{\infty}_{(p,0)}(\Omega),$$

$$AC_p(\Omega) = Z_{(p,0)}(\bar{\Omega}) = A_p(\Omega) \cap C_{(p,0)}(\bar{\Omega}).$$

$A_p(K)$ denotes the inductive limit of the spaces $A_p(\Omega)$, $\Omega \supset K$, where K is a compact set in $\mathbf{C}^n$.

If $\alpha \in C^m_{(p,q)}(\Omega)$ and $\varphi \in C^{\infty}(\mathbf{C}^n)$ with $\operatorname{supp} \varphi \Subset \Omega$, then we shall suppose that $\varphi\alpha$ is extended by zero to $C\Omega$, so that $\varphi\alpha \in C^m_{(p,q)}(\mathbf{C}^n)$.

The index z (or ζ) under a domain of integration signifies that z (respectively ζ) is the variable of integration.

We shall say that D is a *domain with boundary of class* $C^{m,\lambda}$ if $D = \{z \in \mathbf{C}^n : \rho(z) < 0\}$, where $\rho \in C^{m,\lambda}(\mathbf{C}^n)$ and $\operatorname{grad} \rho \neq 0$ on ∂D.

If $m = 1$ and $\lambda = 0$, D is said to have *smooth boundary*.

(²) The bar denotes closure.

CHAPTER I

INTEGRAL REPRESENTATION OF EXTERIOR DIFFERENTIAL FORMS AND ITS IMMEDIATE CONSEQUENCES

§1. The Martinelli-Bochner-Koppelman formula

1°. The definition of the determinant of a numerical matrix uses only the operations of addition and multiplication; hence we can form the determinant of a matrix whose elements lie in an arbitrary ring.

A determinant of order n is the algebraic sum

$$\begin{vmatrix} a_{11} & a_{12} & \cdots & a_{1n} \\ a_{21} & a_{22} & \cdots & a_{2n} \\ \vdots & \vdots & & \vdots \\ a_{n1} & a_{n2} & \cdots & a_{nn} \end{vmatrix} = \sum_I (-1)^{\varepsilon_I} a_{i_1 1} \cdot \cdots \cdot a_{i_n n},$$

where ε_I is the parity of the permutation $I = (i_1, \ldots, i_n)$. Thus the determinant is expanded according to the usual rule, the place where a factor occurs in the summand being determined by the column to which the factor belongs.

In connection with the expansion of a determinant, it is convenient to introduce the following notation.

Let $\theta^1, \ldots, \theta^m$ be n-dimensional vectors. Then

$$D_{\nu_1, \ldots, \nu_m}(\theta^1, \ldots, \theta^m)$$

is the determinant of order $n = \nu_1 + \cdots + \nu_m$ whose first ν_1 columns are θ^1, the next ν_2 columns are θ^2, and so on.

We state some properties([1]) of determinants which will be needed subsequently.

([1]) These properties are trivial from the point of view of algebra; however for some of them we have not been able to find appropriate references, and have taken the liberty of listing them for the convenience of the reader.

PROPERTY 1. *If each element of the kth row (or column) of a determinant has been expressed as the sum of two terms, then the determinant is equal to the sum of two determinants having the same rows (columns) as the given determinant except the kth, which consists of the first and second summands respectively.*

PROPERTY 2. *If two rows of a determinant are interchanged, then the determinant changes sign.*

PROPERTY 3. *If all the elements of a certain column commute with all the elements of an adjacent column, then the interchange of these two columns changes the sign of the determinant; if they anticommute,*(2) *then the determinant is unaffected by the interchange.*

PROPERTY 4. *If all the elements of a column (or row) are multiplied by an element which commutes with all the elements of the determinant, then the determinant also gets multiplied by this element.*

PROPERTY 5. *If all the elements of the first column are multiplied on the left by a certain element, then the determinant also gets multiplied on the left by the same element.*

For the remaining properties which we state, we assume that the entries of the determinant lie in the ring of double forms (see de Rham [1], Chapter II, §7). Observe that Property 4 yields

PROPERTY 4′. *If all the elements of a column (or row) are multiplied by the same function, then the determinant gets multiplied by this function.*

PROPERTY 6. *If the rows of the determinant are linearly dependent,*(3) *then the determinant vanishes.*

The proofs are the same as for the usual determinants.

For forms, and hence also for double forms, we have the following assertion: *If the* α_i, $i = 1,\dots,m$, *are forms of class* C^1, *of degree* q_i *in* z, *then*

$$\bar\partial_z(\alpha_1 \wedge \cdots \wedge \alpha_m) = \sum_{i=1}^{m} (-1)^{q_1+\cdots+q_{i-1}} \alpha_1 \wedge \cdots \wedge \alpha_{i-1} \wedge \bar\partial_z \alpha_i \wedge \alpha_{i+1} \wedge \cdots \wedge \alpha_m.$$

(2) By anticommutation we mean the relation $ab = -ba$.

(3) The rows are said to be *linearly dependent* if some linear combination of them with functions as coefficients vanishes identically, and the coefficients do not have a common zero.

From this we get

PROPERTY 7. *If the elements of the ith columns θ^i of a determinant are C^1 double forms of degree q_i in z, $i = 1,\ldots,n$, then*

$$\bar\partial_z D_{1,\ldots,1}(\theta^1,\ldots,\theta^n) = \sum_{i=1}^{n} (-1)^{q_1+\cdots+q_{i-1}} D_{1,\ldots,1}(\theta^1,\ldots,\theta^{i-1}\bar\partial_z\theta^i, \theta^{i+1},\ldots,\theta^n).$$

Here and in the sequel, $\bar\partial_z\theta^i = (\bar\partial_z\theta_1^i,\ldots,\bar\partial_z\theta_n^i)$.

2°. Let $w(\zeta, z) = (w_1,\ldots,w_n)$ be a vector-valued function defined on $U \subset \mathbf{C}_\zeta^n \times \mathbf{C}_z^n$, with the property that $\langle w, \zeta - z\rangle = 1$. If $w_j \in C^1(U)$ for all j, we let

$$W_{p,q}(w,\zeta,z) = \frac{(-1)^{q+p(n-q-1)}\binom{n-1}{q}}{(2\pi i)^n p!\,(n-p)!} D_{1,q,n-q-1}(w, \bar\partial_z w, \bar\partial_\zeta w) \wedge D_{p,n-p}(\partial z, \partial\zeta).$$

Here the subscripts z and ζ of $\bar\partial$ indicate the group of variables with respect to which $\bar\partial$ is acting, and ∂z and $\partial\zeta$ denote the vector forms $(dz_1,\ldots,dz_n)$ and $(d\zeta_1,\ldots,d\zeta_n)$ respectively.

We also set $W_{p,-1} \equiv W_{p,n} \equiv 0$.

The kernels of the integral representation for exterior differential forms will be the forms $W_{p,q}(w,\zeta,z)$, for a special choice of the vector w, namely $w = t(\zeta,z) = (\bar\zeta - \bar z)/|\zeta - z|^2$. We therefore introduce the notation

$$U_{p,q}(\zeta,z) = W_{p,q}(t(\zeta,z),\zeta,z).$$

Observe that $W_{0,0}(w,\zeta,z)$ is the Cauchy-Fantappiè kernel; for this form we shall also write

$$\omega(w,\zeta,z) = W_{0,0}(w,\zeta,z).$$

We can now state the Martinelli-Bochner formula for exterior differential forms:

Let D be a bounded domain with piecewise smooth boundary, and let $\gamma \in C^1_{(p,q)}(\bar D)$. Then

$$I^1_{p,q}(D,\gamma) - I^2_{p,q}(D,\bar\partial\gamma) - \bar\partial I^2_{p,q-1}(D,\gamma) = \begin{cases}\gamma(z), & z\in D,\\ 0, & z\notin \bar D,\end{cases} \tag{1.1}$$

where

$$I^1_{p,q}(D,\gamma)(z) = \int_{(\partial D)_\zeta} \gamma(\zeta) \wedge U_{p,q}(\zeta, z),$$

$$I^2_{p,q-1}(D,\gamma)(z) = \int_{D_\zeta} \gamma(\zeta) \wedge U_{p,q-1}(\zeta, z).$$

The proof of (1.1) will be given in 5°. This formula is valid in greater generality. For example, we could assume that D is a finite union of domains with piecewise smooth boundary, whose closures are pairwise disjoint.

Also, for $q = n$, (1.1) holds for any open set D and any form $\gamma \in C^1_{(p,n)}(D)$ whose coefficients are absolutely integrable in D. Indeed, let B be a ball with $B \Subset D$. Since (1.1) holds for B, we have for $z \in B$

$$\begin{aligned} \gamma(z) &= \bar\partial \int_{B_\zeta} \gamma(\zeta) \wedge U_{p,n-1}(\zeta, z) \\ &= \bar\partial \int_{D_\zeta} \gamma(\zeta) \wedge U_{p,n-1}(\zeta, z) - \bar\partial \int_{(D\setminus B)_\zeta} \gamma(\zeta) \wedge U_{p,n-1}(\zeta, z). \quad (1.2) \end{aligned}$$

For $z \in B$, we may differentiate the second integral under the integral sign. Now $\bar\partial_z U_{p,n-1} = 0$ by Lemma 1.2; hence (1.2) reduces to (1.1) for $q = n$, under the assumptions made on D and γ.

3°. Observe that

$$\bar\partial_z t = \bar\partial_z \frac{\bar\zeta - \bar z}{|\zeta - z|^2} = (\bar\zeta - \bar z)\bar\partial_z \frac{1}{|\zeta - z|^2} - \frac{1}{|\zeta - z|^2}\bar\partial_z \bar z.$$

Hence for $q \geqslant 1$

$$\begin{aligned} D_{1,q,n-q-1}&\big(t, \bar\partial_z t, \bar\partial_\zeta t\big) \\ &= \frac{1}{|\zeta - z|^2}\bar\partial_z \frac{1}{|\zeta - z|^2} \wedge D_{2,q-1,n-q-1}\big(\bar\zeta - \bar z, \bar\partial_z t, \bar\partial_\zeta t\big) \\ &\quad - \frac{1}{|\zeta - z|^4} D_{1,1,q-1,n-q-1}\big(\bar\zeta - \bar z, \bar\partial_z \bar z, \bar\partial_z t, \bar\partial_\zeta t\big). \end{aligned}$$

The first determinant on the right has two identical columns; hence it vanishes, by Property 3 of determinants.

Repeating this procedure $q - 1$ times more, we get

$$D_{1,q,n-q-1}\big(t, \bar\partial_z t, \bar\partial_\zeta t\big) = \frac{(-1)^q}{|\zeta - z|^{2q+2}} D_{1,q,n-q-1}\big(\bar\zeta - \bar z, \bar\partial_z \bar z, \bar\partial_\zeta t\big).$$

Proceeding analogously with the last $n-q-1$ columns, we get

$$D_{1,q,n-q-1}\left(t, \bar{\partial}_z t, \bar{\partial}_\zeta t\right)$$

$$= \frac{(-1)^q}{|\zeta - z|^{2n}} D_{1,q,n-q-1}\left(\bar{\zeta} - \bar{z}, \bar{\partial}_z \bar{z}, \bar{\partial}_\zeta \bar{\zeta}\right)$$

$$= \frac{(-1)^q q!(n-q-1)!}{|\zeta - z|^{2n}} \sum_I{}' \sum_{\substack{k=1\\k\notin I}}^n \sigma(I,k)\left(\bar{\zeta}_k - \bar{z}_k\right) d\bar{z}_I d\bar{\zeta}[I,k], \quad (1.3)$$

where $I = (i_1, \ldots, i_q)$ is a multi-index, the prime over the summation sign indicates that the summation is over increasing multi-indices, $k \notin I$ means that k is different from $i_1, \ldots, i_q$, and the constant $\sigma(I, k)$ is determined by the equality

$$d\bar{z}_k \wedge d\bar{z}_I \wedge d\bar{z}[I,k] = \sigma(I,k) d\bar{z}.$$

Since

$$D_{p,n-p}(\partial z, \partial\zeta) = p!(n-p)! \sum_J{}' \sigma(J) dz_J \cdot d\zeta[J],$$

we deduce from (1.3) that

$$U_{p,q}(\zeta, z) = \frac{(-1)^{p(n-q-1)}(n-1)!}{(2\pi i)^n} \sum_{I,J}{}' \sum_{\substack{k=1\\k\notin I}}^n \sigma(I,k)\sigma(J)$$

$$\times \frac{\bar{\zeta}_k - \bar{z}_k}{|\zeta - z|^{2n}} d\bar{\zeta}[I,k] \wedge d\zeta[J] d\bar{z}_I \wedge dz_J. \quad (1.4)$$

Here $\sigma(J)$ is determined by

$$dz_J \wedge d\bar{z}[J] = \sigma(J) dz.$$

For the proof of (1.1), we need one more form of expression for $U_{0,q}$. Let

$$V_q = \frac{(-1)^n (n-1)!}{(2\pi i)^n} \sum_I{}' \sum_{\substack{k=1\\k\notin I}}^n \sigma(I,k)(-1)^{k-1} d\bar{\zeta}[I,k] \wedge d\zeta[k] d\bar{z}_I;$$

then

$$U_{0,q} = \frac{(-1)^{q+1}}{1-n} \partial_\zeta \frac{1}{|\zeta - z|^{2n-2}} \wedge V_q. \quad (1.5)$$

4°. Throughout this subsection, the letter w (possibly with superscripts) denotes a vector-valued function $w(\zeta, z) = (w_1(\zeta, z), \ldots, w_n(\zeta, z))$, where $(\zeta, z) \in U \subset (\mathbf{C}^n_\zeta \times \mathbf{C}^n_z) \setminus \{(\zeta, z) \colon \zeta = z\}$, and $\langle w, \zeta - z \rangle = 1$.

LEMMA 1.1. *The determinant*

$$D_{1,\dots,1}\left(w^1, \bar\partial_z w^2, \dots, \bar\partial_z w^l, \bar\partial_\zeta w^{l+1}, \dots, \bar\partial_\zeta w^n\right)$$

remains unaltered if the vector w_1 *is replaced by any other* w *defined in the same domain.*

PROOF. Subtract from the given determinant the new determinant

$$D_{1,\dots,1}\left(w, \bar\partial_z w^2, \dots, \bar\partial_z w^l, \bar\partial_\zeta w^{l+1}, \bar\partial_\zeta w^n\right).$$

By Property 1,

$$D_{1,\dots,1}\left(w^1 - w, \bar\partial_z w^2, \dots, \bar\partial_z w^l, \bar\partial_\zeta w^{l+1}, \dots, \bar\partial_\zeta w^n\right).$$

Since $\langle w^1 - w, \zeta - z\rangle = \langle \bar\partial_z w^i, \zeta - z\rangle = \langle \bar\partial_\zeta w^i, \zeta - z\rangle = 0$, we see by Property 6 that this last determinant vanishes.

LEMMA 1.2. *If* w *is of class* C^2 *and* $0 \leqslant q \leqslant n$, *then*

$$\bar\partial_\zeta W_{p,q}(w, \zeta, z) = (-1)^{p+q} \bar\partial_z W_{p,q-1}(w, \zeta, z). \tag{1.6}$$

In particular, $\bar\partial_\zeta W_{p,0} = \bar\partial_z W_{p,n-1} = 0$.

PROOF. For $1 \leqslant q \leqslant n$, we have by Property 7

$$\begin{aligned}\bar\partial_z D_{1,q-1,n-q}\left(w, \bar\partial_z w, \bar\partial_\zeta w\right) &= D_{q,n-q}\left(\bar\partial_z w, \bar\partial_\zeta w\right) \\ &\quad + (-1)^{q-1}(n-q) D_{1,q-1,1,n-q-1}\left(w, \bar\partial_z w, \bar\partial_z \bar\partial_\zeta w, \bar\partial_\zeta w\right).\end{aligned} \tag{1.7}$$

For $0 \leqslant q \leqslant n-1$, the same property yields

$$\begin{aligned}\bar\partial_\zeta D_{1,q,n-q-1}\left(w, \bar\partial_z w, \bar\partial_\zeta w\right) &= (-1)^q D_{q,n-q}\left(\bar\partial_z w, \bar\partial_\zeta w\right) \\ &\quad + q D_{1,q-1,1,n-q-1}\left(w, \bar\partial_z w, \bar\partial_\zeta \bar\partial_z w, \bar\partial_\zeta w\right).\end{aligned} \tag{1.8}$$

By Property 6, the first determinant in each of (1.7) and (1.8) vanishes. For $q = n$ in (1.7), and for $q = 0$ in (1.8), the second term vanishes; hence from the definition of $W_{p,n-1}$ and $W_{p,n}$ we see that (1.6) holds for $q = 0$ and n.

For $1 \leqslant q \leqslant n-1$, the second determinants in (1.7) and (1.8) are equal to each other, since $\bar\partial_z \bar\partial_\zeta w = \bar\partial_\zeta \bar\partial_z w$. Here $\bar\partial_z \bar\partial_\zeta w$ is regarded as a double form. Therefore

$$\bar\partial_\zeta\left[(n-q) D_{1,q,n-q-1}\left(w, \bar\partial_z w, \bar\partial_\zeta w\right)\right] = \bar\partial_z\left[(-1)^{q-1} q D_{1,q-1,n-q}\left(w, \bar\partial_z w, \bar\partial_\zeta w\right)\right].$$

Multiplying this equality by

$$\frac{(n-1)!\,(-1)^{p(n-q-1)}}{(2\pi i)^n q!\,(n-q)!\,p!\,(n-p)!} D_{p,n-p}(\partial z, \partial \zeta),$$

we obtain (1.6).

LEMMA 1.3. *If w^1 and w^2 are vector-valued functions of class C^2 in ζ, then*

$$W_{p,0}(w^2, \zeta, z) - W_{p,0}(w^1, \zeta, z) = \bar{\partial}_\zeta \mu_0 \wedge \frac{1}{p!\,(n-p)!} \times D_{p,n-p}(\partial z, \partial \zeta),$$

where

$$\mu_0 = \frac{(n-1)!}{(2\pi i)^n} \sum_{j=0}^{n-2} D_{1,1,j,n-j-2}(w^1, w^2, \bar{\partial}_\zeta w^1, \bar{\partial}_\zeta w^2).$$

In particular,

$$\omega(w^2, \zeta, z) - \omega(w^1, \zeta, z) = \bar{\partial}_\zeta \mu_0 \wedge d\zeta.$$

The proof is a verification, using Properties 3 and 7 and Lemma 1.1. Similar considerations are used in the proof of the next lemma.

LEMMA 1.4. *If $U \subset \mathbf{C}^n_\zeta \times \mathbf{C}^n_z$ is such that there exists a vector-valued function $u(\zeta, z) \in C^2(U)$ holomorphic in z for fixed ζ and satisfying $\langle u, \zeta - z \rangle = 1$, then for all $w \in C^2(U)$*

$$W_{p,q}(w, \zeta, z) = (\bar{\partial}_z \mu^1_{q-1} + \bar{\partial}_\zeta \mu^2_q) \wedge \frac{1}{p!\,(n-p)!} D_{p,n-p}(\partial z, \partial \zeta), \quad (1.9)$$

where μ^1_q and μ^2_q are forms which are linear combinations of forms of the type

$$D_{1,1,q,r,n-q-r-2}(w, u, \bar{\partial}_z w, \bar{\partial}_\zeta u, \bar{\partial}_\zeta w). \quad (1.10)$$

PROOF. We first show that

$$D_{1,q,n-q-1}(w, \bar{\partial}_z w, \bar{\partial}_\zeta u) = \bar{\partial}_z \alpha^{q-1}_{n-q-1} \quad (1.11)$$

and that for $0 \leqslant r \leqslant n-q-2$ there exist constants c_r such that

$$\begin{aligned} D_{1,q,r,n-q-r-1}(w, \bar{\partial}_z w, \bar{\partial}_\zeta u, \bar{\partial}_\zeta w) - c_r D_{1,q,r+1,n-q-r-2}(w, \bar{\partial}_z w, \bar{\partial}_\zeta u, \bar{\partial}_\zeta w) \\ = \bar{\partial}_z \alpha^{q-1}_r + \bar{\partial}_\zeta \beta^q_r, \end{aligned} \quad (1.12)$$

where α^q_r and β^q_r are forms differing from (1.10) by a constant factor.

Since u, and hence $\bar{\partial}_\zeta u$, is holomorphic in z, by Property 7 we have

$$\begin{aligned} &\bar{\partial}_z D_{1,1,q-1,r,n-q-r-1}(u, w, \bar{\partial}_z w, \bar{\partial}_\zeta u, \bar{\partial}_\zeta w) \\ &\quad = D_{1,q,r,n-q-r-1}(u, \bar{\partial}_z w, \bar{\partial}_\zeta u, \bar{\partial}_\zeta w) \\ &\quad + (-1)^{q-1}(n-q-r-1) D_{1,1,q-1,r,1,n-q-r-2} \\ &\quad \times (u, w, \bar{\partial}_z w, \bar{\partial}_\zeta u, \bar{\partial}_z \bar{\partial}_\zeta w, \bar{\partial}_\zeta w). \end{aligned} \quad (1.13)$$

For $r = n - q - 1$, the second term on the right side of (1.3) disappears, and the first term is equal to the left side of (1.11) by Lemma 1.1; hence (1.11) holds.

For $0 \leqslant r \leqslant n - q - 2$, we have

$$\begin{aligned}
\bar{\partial}_\zeta D_{1,1,q,r,n-q-r-2}&\left(u, w, \bar{\partial}_z w, \bar{\partial}_\zeta u, \bar{\partial}_\zeta w\right) \\
&= (-1)^{q-1} D_{1,q,r+1,n-q-r-2}\left(w, \bar{\partial}_z w, \bar{\partial}_\zeta u, \bar{\partial}_\zeta w\right) \\
&\quad + (-1)^q D_{q,q,r,n-q-r-1}\left(u, \bar{\partial}_z w, \bar{\partial}_\zeta u, \bar{\partial}_\zeta w\right) \\
&\quad + q D_{1,1,q-1,1,r,n-q-r-2}\left(u, w, \bar{\partial}_z w, \bar{\partial}_\zeta \bar{\partial}_z w, \bar{\partial}_\zeta u, \bar{\partial}_\zeta w\right).
\end{aligned}$$

Hence by (1.13) and Lemma 1.1 we get

$$\begin{aligned}
\bar{\partial}_z q D_{1,1,q-1,r,n-q-r-1}&\left(u, w, \bar{\partial}_z w, \bar{\partial}_\zeta u, \bar{\partial}_\zeta w\right) \\
&+ \bar{\partial}_\zeta (-1)^q (n - q - r - 1) D_{1,1,q,r,n-q-r-2}\left(u, w, \bar{\partial}_z w, \bar{\partial}_\zeta u, \bar{\partial}_\zeta w\right) \\
= (n - r - 1)& D_{1,q,r,n-q-r-1}\left(w, \bar{\partial}_z w, \bar{\partial}_\zeta u, \bar{\partial}_\zeta w\right) \\
- (n - q - r - 1)& D_{1,q,r+1,n-q-r-2}\left(w, \bar{\partial}_z w, \bar{\partial}_\zeta u, \bar{\partial}_\zeta w\right).
\end{aligned}$$

Dividing both sides by $n - r - 1$, we obtain (1.12).

To prove (1.9), we now need only multiply (1.12) by $c_0 \cdot \cdots \cdot c_{r-1}$ and sum over r, and then add (1.11) multiplied by $c_0 \cdot \cdots \cdot c_{n-q-2}$.

LEMMA 1.5.

$$U_{p,q}(\zeta, z) = -U_{n-p,n-q-1}(z, \zeta).$$

PROOF. We have

$$\begin{aligned}
U_{p,q}(\zeta, z) &= \frac{(-1)^{q+p(n-q-1)}\binom{n-1}{q}}{(2\pi i)^n p!\,(n-p)!} \\
&\quad \times D_{1,q,n-q-1}\left(t(\zeta, z), \bar{\partial}_z t(\zeta, z), \bar{\partial}_\zeta t(\zeta, z)\right) \wedge D_{p,n-p}(\partial z, \partial \zeta), \\
U_{n-p,n-q-1}(z, \zeta) &= \frac{(-1)^{n-q,-1+(n-p)q}\binom{n-1}{q}}{(2\pi i)^n p!(n-p)!} \\
&\quad \times D_{1,n-q-1,q}\left(t(z, \zeta), \partial_\zeta t(z, \zeta), \bar{\partial}_z t(z, \zeta)\right) \wedge D_{n-p,p}(\partial \zeta, \partial z).
\end{aligned}$$

The required result now follows by Property 3, if we take into account the relation $t(\zeta, z) = -t(z, \zeta)$.

LEMMA 1.6.

$$\omega(w, \zeta, z) = \frac{(n-1)!}{(2\pi i)^n} \sum_{k=1}^{n} (-1)^{k-1} w_k dw[k] \wedge d\zeta$$
$$= \frac{(n-1)!}{(2\pi i)^n} \cdot \frac{(-1)^{k-1} dw[k] \wedge d\zeta}{\zeta_k - z_k},$$

where the second equality holds for $\zeta_k \neq z_k$.

PROOF. To prove the first equality, we need only expand the determinant $D_{1,n-1}(w, \bar{\partial}_\zeta w)$ by elements of the first column and observe that $D_n(\partial\zeta) = n! d\zeta$. To prove the second equality, we use Lemma 1.1 to replace the first column of $D_{1,n-1}(w, \bar{\partial}_\zeta w)$ by the column whose kth component is $(\zeta_k - z_k)^{-1}$ and all other components are zero, and expand the resulting determinant according to elements of this column.

5°. PROOF OF (1.1). We need the following lemma.

LEMMA 1.7. *If* $f \in C^1$ *in a neighborhood of* $0 \in \mathbf{C}^n$, *then*

$$\lim_{\varepsilon \to 0} \int_{|\zeta|=\varepsilon} \frac{f(\zeta) d\bar{\zeta} \wedge d\zeta[k]}{|\zeta|^{2n-m}} = \begin{cases} (-1)^{n-k-1} \dfrac{(2\pi i)^n}{n!} \dfrac{\partial f}{\partial \bar{\zeta}_k}(0); & m = 0, \\ 0; \quad m > 0. \end{cases} \tag{1.14}$$

In particular,

$$\lim_{\varepsilon \to 0} \int_{|\zeta|=\varepsilon} \frac{g(\zeta) \zeta_j d\bar{\zeta} \wedge d\zeta[k]}{|\zeta|^{2n}} = (-1)^{n-k-1} \frac{(2\pi i)^n}{n!} g(0) \cdot \delta_{jk}.$$

Here δ_{jk} *is the Kronecker symbol* ($\delta_{jk} = 0$ *for* $j \neq k$, *and* $\delta_{jj} = 1$).

PROOF. We have

$$\int_{|\zeta|=\varepsilon} \frac{f(\zeta) d\bar{\zeta} \wedge d\zeta[k]}{|\zeta|^{2n-m}} = \frac{1}{\varepsilon^{2n-m}} \int_{|\zeta|=\varepsilon} f(\zeta) d\bar{\zeta} \wedge d\zeta[k]$$
$$= (-1)^{n-k-1} \cdot \varepsilon^{m-2n} \int_{|\zeta|<\varepsilon} \frac{\partial f}{\partial \bar{\zeta}_k} d\bar{\zeta} \wedge d\zeta$$
$$= (-1)^{n-k-1} \cdot \varepsilon^{m-2n} \left[\frac{\partial f}{\partial \bar{\zeta}_k}(0) \int_{|\zeta|<\varepsilon} d\bar{\zeta} \wedge d\zeta + \int_{|\zeta|<\varepsilon} \left(\frac{\partial f}{\partial \bar{\zeta}_k}(\zeta) - \frac{\partial f}{\partial \bar{\zeta}_k}(0) \right) d\bar{\zeta} \wedge d\zeta \right]. \tag{1.15}$$

The formula for the volume of a ball yields

$$\int_{|\zeta|<\varepsilon} d\bar{\zeta} \wedge d\zeta = \frac{(2\pi i)^n}{n!} \varepsilon^{2n},$$

so that the first term on the right side of (1.15) gives the right side of (1.14). But the limit of the second term as $\varepsilon \to 0$ is zero, since the integrand tends to zero as $\varepsilon \to 0$. So the first part of the lemma is proved.

To prove the second part, we need only observe that

$$\frac{\partial}{\partial \zeta_k}(g\zeta_j)(0) = g(0)\delta_{jk}.$$

We now proceed to the proof of (1.1). In view of the additivity of the operators on the left side of (1.1), we need only prove the formula for forms of the type

$$\gamma = f(z)\,d\bar{z}_I \wedge dz_J.$$

Here $f \in C^1(\bar{D})$, $I = (i_1, \dots, i_q)$ and $J = (j_1, \dots, j_p)$.

Now

$$\begin{aligned}\gamma(\zeta) \wedge U_{p,q}(\zeta, z) &= \frac{(-1)^{q+p(n-q-1)}\binom{n-1}{q}}{(2\pi i)^n p!\,(n-p)!} f(\zeta)\,d\bar{\zeta}_I \wedge d\zeta_J \\ &\quad \wedge D_{1,q,n-q-1}\big(t, \bar{\partial}_z t, \bar{\partial}_\zeta t\big) \wedge D_{p,n-p}(\partial z, \partial \zeta) \\ &= f(\zeta)\,d\bar{\zeta}_I \wedge \frac{(-1)^q\binom{n-1}{q}}{(2\pi i)^n} D_{1,q,n-q-1}\big(t, \bar{\partial}_z t, \bar{\partial}_\zeta t\big) \wedge d\zeta \wedge dz_J \\ &= f(\zeta)\,d\bar{\zeta}_I \wedge U_{0,q}(\zeta, z) \wedge dz_J,\end{aligned}$$

and analogously

$$\bar{\partial}\gamma \wedge U_{p,q}(\zeta, z) = \bar{\partial}\big(f(\zeta)\,d\bar{\zeta}_I\big) \wedge U_{0,q} \wedge dz_J,$$

$$\gamma \wedge U_{p,q-1}(\zeta, z) = f(\zeta)\,d\bar{\zeta}_I \wedge U_{0,q-1} \wedge dz_J.$$

These equalities show that (1.1) needs only to be proved for forms of type $(0, q)$, or more precisely for forms of the type $\gamma = f(z)\,d\bar{z}_I$.

Finally, the kernels $U_{0,q}$ are invariant under the transformations $\zeta_j \to \zeta_{\tau(j)}$, $z_j \to z_{\tau(j)}$, $j = 1, \dots, n$, τ being an arbitrary permutation. Indeed, under such a transformation, the same rows of the two determinants occurring in the expression for $U_{0,q}$ change their places. Thus it suffices to prove (1.1) for the forms of the type

$$\gamma = f(z)\,d\bar{z}_1 \wedge \cdots \wedge d\bar{z}_q.$$

For $z \notin D$, (1.1) is an immediate consequence of Stokes' formula and Lemma 1.2.

Let $z \in D$. We set

$$g = (1-n)^{-1}|\zeta - z|^{2-2n}, \qquad \theta_q^* = U_{0,q} - \bar\partial_z g \wedge V_{q-1}.$$

Since $\gamma \wedge U_{0,q}$ has type $(n, n\ \ 1)$ in ζ, and $\gamma \wedge \bar\partial_z g \wedge V_{q-1}$ is of type $(n-1, n)$ in ζ, we have

$$\begin{aligned} d\zeta(\gamma \wedge \theta_q^*) &= \bar\partial_\zeta(\gamma \wedge U_{0,q}) - \partial_\zeta(\gamma \wedge \bar\partial_z g \wedge V_{q-1}) \\ &= \bar\partial_\zeta\gamma \wedge U_{0,q} + (-1)^q \gamma \wedge \bar\partial_\zeta U_{0,q} - \partial_\zeta\gamma \wedge \bar\partial_z g \wedge V_{q-1} \\ &\quad - (-1)^q \gamma \wedge \partial_\zeta\bar\partial_z g \wedge V_{q-1}. \end{aligned}$$

As we already observed, $\bar\partial_z\partial_\zeta g = \partial_\zeta\bar\partial_z g$; hence it follows from (1.5) and Lemma 1.2 that

$$\begin{aligned} (-1)^q \gamma \wedge \partial_\zeta\bar\partial_z g \wedge V_{q-1} &= (-1)^q \gamma \wedge \bar\partial_z(\partial_\zeta g \wedge V_{q-1}) \\ &= \gamma \wedge \bar\partial_z U_{0,q-1} = (-1)^q \gamma \wedge \bar\partial_\zeta U_{0,q}. \end{aligned}$$

This together with the preceding relation yields

$$d_\zeta(\gamma \wedge \theta_q^*) = \bar\partial_\zeta\gamma \wedge U_{0,q} - \partial_\zeta\gamma \wedge \bar\partial_z g \wedge V_{q-1}. \tag{1.16}$$

Therefore the second integral in

$$i(z) = \int_{\partial D_\zeta} \gamma \wedge \theta_q^* - \int_{D_\zeta} d_\zeta(\gamma(\zeta) \wedge \theta_q^*) \tag{1.17}$$

is absolutely convergent.

We shall now show that $i(z)$ is precisely the left-hand side of (1.1). For this we consider the form $\gamma \wedge gV_{q-1}$. This is of type $(n-1, n)$ in ζ; hence by Stokes' theorem

$$\begin{aligned} &\int_{\partial D_\zeta} \gamma \wedge gV_{q-1} - \int_{(|\zeta - z| = \varepsilon)_\zeta} \gamma \wedge gV_{q-1} \\ &\qquad = \int_{(D_\varepsilon)_\zeta} \partial\gamma \wedge gV_{q-1} + (-1)^q \int_{(D_\varepsilon)_\zeta} \gamma \wedge \partial_\zeta g \wedge V_{q-1}, \end{aligned} \tag{1.18}$$

where $D_\varepsilon = \{\zeta \in D \mid |\zeta - z| > \varepsilon\}$. Lemma 1.7 shows that the second integral on the right in (1.8) tends to zero as $\varepsilon \to 0$, so that (1.5) and (1.18) yield

$$\int_{D_\zeta} \gamma \wedge U_{0,q-1} = \int_{\partial D_\zeta} \gamma(\zeta) \wedge gV_{q-1} - \int_{D_\zeta} \partial\gamma(\zeta) \wedge gV_{q-1}. \tag{1.19}$$

In both the integrals on the right side of (1.19), we may differentiate under the integral sign; the first is a proper integral, while the second has a singularity of the form $|z - \zeta|^{1-2n}$ after differentiating the integrand and hence is absolutely convergent. Hence, the left side of (1.19) is likewise differentiable, and we get

$$\int_{\partial D_\zeta} \gamma(\zeta) \wedge \bar\partial_z g \wedge V_{q-1} - \int_{D_\zeta} \partial\gamma(\zeta) \wedge \bar\partial_z g \wedge V_{q-1} - \bar\partial \int_{D_\zeta} \gamma \wedge U_{0,q-1} = 0. \tag{1.20}$$

Adding (1.17) and (1.20), and using the definition of $U_{0,q}$ and (1.16), we obtain

$$i(z) = \int_{\partial D_\zeta} \gamma(\zeta) \wedge U_{0,q} - \int_{D_\zeta} \bar\partial\gamma \wedge U_{0,q} - \bar\partial \int_{D_\zeta} \gamma \wedge U_{0,q-1}.$$

We shall show that $i(z) = \gamma(z)$. From (1.17) and Stokes' theorem applied to D_ε we get

$$i(z) = \lim_{\varepsilon \to 0} \int_{|\zeta - z| = \varepsilon} \gamma(\zeta) \wedge \theta_q^*.$$

Let us compute this limit. To do this, we go over from θ_q^* to the form

$$\theta_q = (-1)^q \bar\partial_\zeta g \wedge V_q - \bar\partial_z g \wedge V_{q-1}.$$

From the invariance of the first differential, it follows that $d_\zeta g|_K = 0$, where $K = \{\zeta \in \mathbf{C}^n \colon |\zeta - z| = \varepsilon\}$, or $\bar\partial_\zeta g|_K = \partial_\zeta g|_K$; hence (1.5) and the definition of θ_q^* show that $\theta_q^*|_K = \theta_q|_K$.

We set

$$g_k = \frac{\partial g}{\partial \bar\zeta_k} = -\frac{\partial g}{\partial \bar z_k} = \frac{\zeta_k - z_k}{|\zeta - z|^{2n}}.$$

Then

$$\gamma(\zeta) \wedge \theta_q = (-1)^n \frac{(n-1)!}{(2\pi i)^n} f(\zeta)$$
$$\times \Big\{ (-1)^q d\bar\zeta_1 \wedge \cdots \wedge d\bar\zeta_q \wedge \sum_{k=1}^n g_k d\bar\zeta_k \wedge {\sum_I}' \sum_{j \notin I} \sigma(I, j)(-1)^{j-1} d\bar\zeta\,[I, j]$$
$$\wedge d\zeta[j] d\bar z_I - d\bar\zeta_1 \wedge \cdots \wedge d\bar\zeta_q \wedge \sum_{j=1}^n (-g_k) d\bar z_k$$
$$\wedge {\sum_J}' \sum_{k \notin J} \sigma(J, k)(-1)^{k-1} \times d\bar\zeta\,[J, k] \wedge d\zeta[k] d\bar z_J \Big\}$$

$$= \frac{(-1)^n (n-1)!}{(2\pi i)^n} f(\zeta) \Big\{ (-1)^q \sum_{k=q+1}^{n} g_k d\bar{\zeta}_1 \wedge \cdots \wedge d\bar{\zeta}_q \wedge d\bar{\zeta}_k \wedge d\bar{\zeta}\,[1,\ldots,q,k]$$

$$\wedge \Bigg[\sum_{j=1}^{q} \sigma((1,\ldots,[j],\ldots,q,k),j)(-1)^{j-1} d\zeta[j] \wedge d\bar{z}_1 \wedge \cdots [j]$$

$$\cdots d\bar{z}_q \wedge d\bar{z}_k + \sigma((1,\ldots,q),k)(-1)^{k-1} d\zeta[k] \wedge d\bar{z}_1 \wedge \cdots \wedge d\bar{z}_q \Bigg]$$

$$+ d\bar{\zeta}_1 \wedge \cdots \wedge d\bar{\zeta}_q \wedge d\bar{\zeta}\,[1,\ldots,q] \wedge \sum_{k=1}^{q} \sigma((1,\ldots,[k],\ldots,q),k)$$

$$\times (-1)^{k-1} \wedge d\zeta[k] \sum_{j=1}^{n} g_j d\bar{z}_j \wedge d\bar{z}_1 \wedge \cdots [k] \cdots \wedge d\bar{z}_q \Big\}$$

Observe also that

$$\begin{aligned}
&(-1)^q \sigma((1,\ldots,q),k) d\bar{\zeta}_1 \wedge \cdots \wedge d\bar{\zeta}_q \wedge d\bar{\zeta}_k \wedge d\bar{\zeta}\,[1,\ldots,q,k] \\
&= \sigma((1,\ldots,q),k) d\bar{\zeta}_k \wedge \left(d\bar{\zeta}_1 \wedge d\bar{\zeta}_2 \wedge \cdots \wedge d\bar{\zeta}_q\right) \wedge d\bar{\zeta}\,[1,\ldots,q,k] \\
&= d\bar{\zeta}_r\, \sigma((1,\ldots,[k],\ldots,q),k) d\bar{\zeta}_1 \wedge \cdots \wedge d\bar{\zeta}_q \\
&\quad \wedge d\bar{\zeta}\,[1,\ldots,q] d\bar{z}_k \wedge d\bar{z}_1 \wedge \cdots [k] \cdots \wedge d\bar{z}_q \\
&= \sigma((1,\ldots,[k],\ldots,q),k) d\bar{\zeta}_k \\
&\quad \wedge d\bar{\zeta}_1 \wedge \cdots [k] \cdots \wedge d\bar{\zeta}_q \wedge d\bar{\zeta}\,[1,\ldots,q] d\bar{z}_1 \wedge \cdots \wedge d\bar{z}_q = d\bar{\zeta} d\bar{z}_1 \wedge \cdots \wedge d\bar{z}_q.
\end{aligned}$$

Using these equalities, we get

$$\gamma \wedge \theta_q = \frac{(-1)^n (n-1)!}{(2\pi i)^n} f(\zeta) \sum_{k=1}^{n} (-1)^{k-1} g_k d\bar{\zeta} \wedge d\zeta[k] \wedge d\bar{z}_1 \wedge \cdots \wedge d\bar{z}_q + \chi.$$

Here the form χ is a sum of terms of the type

$$\text{const}\, f(\zeta) g_k d\bar{\zeta} \wedge d\zeta[j] \wedge d\bar{z}_I, \qquad j \neq k,\ I = (i_1,\ldots,i_q).$$

Thus finally we get by Lemma 1.7

$$i(z)=\lim_{\varepsilon\to 0}\int_{|\zeta-z|=\varepsilon}\gamma(\zeta)\wedge\theta_q^*=\lim_{\varepsilon\to 0}\int_{|\zeta-z|=\varepsilon}\gamma(\zeta)\wedge\theta_q$$

$$=\frac{(n-1)!}{(2\pi i)^n}\sum_{k=1}^{n}\lim_{\varepsilon\to 0}(-1)^{n+k-1}\int_{|\zeta-z|=\varepsilon}\frac{f(\zeta)(\bar\zeta_k-\bar z_k)}{|\zeta-z|^{2n}}d\bar\zeta$$

$$\wedge d\zeta[k]\wedge d\bar z_1\wedge\cdots\wedge d\bar z_q+\lim_{\varepsilon\to 0}\int_{|\zeta-z|=\varepsilon}\chi$$

$$=\frac{(n-1)!}{(2\pi i)^n}\sum_{k=1}^{n}\frac{(2\pi i)^n}{n!}f(z)\,d\bar z_1\wedge\cdots\wedge d\bar z_q=\gamma(z).$$

§2. Theorems on the saltus of forms

1°. In the theory of functions of one complex variable, the following results are well known (see, for example, Muskelishvili [1], §22).

PROPOSITION 2.1. *Let D be a domain in $\mathbf{C}^1$ with smooth boundary, and let $\varphi\in C^{0,\lambda}(\partial D)$. Then the functions*

$$\varphi^{\pm}(z)=\frac{1}{2\pi i}\int_{\partial D}\frac{\varphi(\zeta)\,d\zeta}{\zeta-z}$$

(φ^+ is defined in D, and φ^- in $C\overline{D}$) can be extended continuously to the closure of the open set on which they are defined, and

$$\varphi^{+}(\zeta)-\varphi^{-}(\zeta)=\varphi(\zeta),\qquad \zeta\in\partial D. \tag{2.1}$$

To formulate the second proposition, we need some notation.

Let D be a domain with smooth boundary, and let $\varphi\in C(\partial D)$ and $z_0\in\partial D$. On a line ∇ passing through z_0 and not tangential to ∂D, choose points $z\in D$ and $z'\notin\overline{D}$ at equal distances from z_0, and consider the limit

$$\lim_{z,\,z'\to z_0}[\varphi^{+}(z)-\varphi^{-}(z')] \tag{2.2}$$

($\varphi^{\pm}$ are defined as in Proposition 2.1).

PROPOSITION 2.2. *The limit (2.2) exists and equals $\varphi(z_0)$. The limit is approached uniformly if the nonobtuse angle between the line ∇ and the tangent to ∂D at z_0 is not less than a fixed $\beta_0>0$.*

It must be noted that Proposition 2.1 solves the so-called *additive Riemann problem*:

Let D be a domain with smooth boundary, and let $\varphi\in C^{0,\lambda}(\partial D)$. Find functions $\varphi^+\in AC(D)$ and $\varphi^-\in AC(C\overline{D})$ such that $\varphi^+-\varphi^-=\varphi$ on D.

We shall need analogues of the Riemann problem and Propositions 2.1 and 2.2 for forms of several complex variables.

2°. Before proceeding to consider the analogues mentioned above, we state the following result without proof.

THEOREM 2.3. *Suppose $\partial D \in C^{m+1,\lambda}$ and $f \in C^{m,\lambda}(\bar{D})$, where $m \geq 0$ and $0 < \lambda \leq 1$. Then the functions defined by the integral*

$$\int_{\partial D_\zeta} \frac{f(\zeta)d\bar{\zeta}\,[k] \wedge d\zeta}{|\zeta - z|^{2n-2}} \tag{2.3a}$$

inside and outside D can be extended to the closure of the respective open sets as functions of class $C^{m+1,\lambda'}$, and the functions defined by the integral

$$\int_{D_\zeta} \frac{f(\zeta)d\bar{\zeta} \wedge d\zeta}{|\zeta - z|^{2n-2}}, \tag{2.3b}$$

as functions of class $C^{m+2,\lambda'}$, $0 < \lambda' < \lambda$.

The proof of this theorem can be carried out along the lines of Gyunter's proof ([1], Chapter II, §19) for the case of real dimension three. But the computations are much more complicated in our case.

Since the coefficients of the kernels $U_{p,q}$ are derivatives of the function $C|\zeta - z|^{2-2n}$, and we may differentiate under the integral sign in (2.3a) and (2.3b), we obtain

COROLLARY 2.4. *If $\partial D \in C^{m+1,\lambda}$ and $\gamma \in C^{m,\lambda}_{(p,q)}(\bar{D})$, then the forms defined by the integral $I^1_{(p,q)}(D, \gamma)$ inside and outside D extend to the closure of the respective open sets as forms of class $C^{m,\lambda'}_{(p,q)}$, and the forms defined by $I^2_{(p,q-1)}(D, \gamma)$ extend as forms of class $C^{m+1,\lambda'}_{(p,q-1)}$, with $0 < \lambda' < \lambda$.*

3°. We observed that (1.1) is the analogue of the Cauchy integral formula for forms of several complex variables. Therefore it is natural to consider an analogue of the integral of Cauchy type an integral of the form $I^1_{p,q}(D, \gamma)(z)$. Here D is a domain with piecewise smooth boundary, and $\gamma \in C_{(p,q)}(\partial D)$. Let γ^+ denote the form defined in D by the above integral and γ^- the one defined in $C\bar{D}$.

THEOREM 2.5. *If $\partial D \in C^2$ and $\gamma \in C^1_{(n,q)}(\partial D)$, $0 \leq q \leq n-1$, then*

$$\gamma = \gamma^+|_{\partial D} - \gamma^-|_{\partial D}$$

(*by Corollary* 2.4, $\gamma^+|_{\partial D}$ *and* $\gamma^-|_{\partial D}$ *are defined*).

PROOF. It suffices to prove that

$$\int_{\partial D} \gamma \wedge \varphi = \int_{\partial D} \left(\gamma^+|_{\partial D} - \gamma^-|_{\partial D}\right) \wedge \varphi \tag{2.4}$$

for any $\varphi \in C^1_{n-q-1}(\partial D)$. We extend φ to $\mathbf{C}^n$ as an $(n-q-1)$-form of class C^1. Then φ can be represented in the form

$$\varphi = \sum_{i=0}^{n-q-1} \varphi_i, \qquad \varphi_i \in C^1_{(i,n-q-i-1)}(\mathbf{C}^n).$$

The forms $\gamma^\pm$ and γ are of type (n, q); therefore $\gamma \wedge \varphi_i = \gamma^\pm \wedge \varphi_i = 0$ for $i = 1, \ldots, n-q-1$. Hence we need prove (2.4) only for $\varphi \in C^1_{(0,n-q-1)}(\mathbf{C}^n)$.

Let ρ be a function defining D; thus, $D = \{z \in \mathbf{C}^n \colon \rho(z) < 0\}$, $\rho \in C^2(\mathbf{C}^n)$, $\operatorname{grad} \rho \neq 0$ on ∂D. Set $D_l = \{z \in \mathbf{C}^n \colon \rho(z) + l^{-1} < 0\}$ and $D^l = D_{-l}$. Then for sufficiently large l we have

$$D_l \Subset D \Subset D^l \Subset \mathbf{C}^n, \qquad \partial D_l \in C^2, \qquad \partial D^l \in C^2,$$

and also, for any $\chi \in C_{2n-1}(\mathbf{C}^n)$,

$$\int_{\partial D} \chi = \lim_{l\to\infty} \int_{\partial D_l} \chi = \lim_{l\to\infty} \int_{\partial D^l} \chi.$$

Thus

$$\begin{aligned}
\int_{\partial D} (\gamma^= - \gamma^-) \wedge \varphi &= \lim_{l\to\infty} \left[\int_{\partial D_l} \gamma^+ \wedge \varphi - \int_{\partial D^l} \gamma^- \wedge \varphi\right] \\
&= \lim_{l\to\infty} \left[\int_{(\partial D_l)_z} \varphi(z) \wedge \int_{\partial D_\zeta} \gamma(\zeta) \wedge U_{n,q}(\zeta, z) \right.\\
&\qquad \left. - \int_{(\partial D^l)_z} \varphi(z) \wedge \int_{\partial D_\zeta} \gamma(\zeta) \wedge U_{n,q}(\zeta, z)\right] \\
&= \lim_{l\to\infty} \int_{\partial D_\zeta} \gamma(\zeta) \wedge \left[\int_{(\partial D^l)_z} \varphi(z) \wedge \left(-U_{n,q}(\zeta, z)\right) \right.\\
&\qquad \left. - \int_{(\partial D_l)_z} \varphi(z) \wedge \left(-U_{n,q}(\zeta, z)\right)\right].
\end{aligned}$$

The reversal of the order of integration is permissible since the integrands are jointly continuous. By Lemma 1.5, $U_{n,q}(\zeta, z) = -U_{0,n-q-1}(z, \zeta)$. Hence by

(1.1) and Stokes' theorem we get

$$\int_{\partial D}(\gamma^+ - \gamma^-)\wedge\varphi = \lim_{l\to\infty}\int_{(\partial D)_\zeta}\gamma(\zeta)\wedge\int_{[\partial(D'\setminus D_l)]_z}\varphi(z)\wedge U_{0,n-q-1}(z,\zeta)$$

$$= \lim_{l\to\infty}\int_{(\partial D)_\zeta}\gamma(\zeta)\wedge\Big\{\varphi(\zeta) + \int_{(D'\setminus D_l)_z}\bar\partial\varphi(z)\wedge U_{0,n-q-1}(z,\zeta)$$

$$+\bar\partial\int_{(D'\setminus D_l)_z}\varphi(z)\wedge U_{0,n-q-2}(z,\zeta)\Big\}$$

$$= \int_{\partial D}\gamma\wedge\varphi + \lim_{l\to\infty}\int_{\partial D_\zeta}\Big\{\gamma(\zeta)\wedge\int_{(D'\setminus D_l)_z}\bar\partial\varphi(z)\wedge U_{0,n-q-1}(z,\zeta)$$

$$+ (-1)^{n+q}\,\bar\partial\gamma(\zeta)\wedge\int_{(D'\setminus D_l)_z}\varphi(z)\wedge U_{0,n-q-2}(z,\zeta)\Big\}.$$

From the form of the kernels $U_{0,r}$, we see that the last limit above is zero as a consequence of

$$\lim_{l\to\infty}\int_{\partial D_\zeta} d\bar\zeta\,[j]\wedge d\zeta\int_{(D'\setminus D_l)_z}\frac{f(\zeta,z)(\bar\zeta_k - \bar z_k)}{|\zeta - z|^{2n}}\,d\bar z\wedge dz = 0, \qquad (2.5)$$

where $f(\zeta, z)\in C(\partial D\times \bar D'\setminus D_l)$.

We show that the integral

$$\int_{\partial D\times(D'\setminus D_l)}\frac{1}{|\zeta - z|^{2n-1}}\,ds\,dv \qquad (2.6)$$

exists. Here dv is the $2n$-dimensional volume element, and ds the area element on ∂D. Since the integrand is positive, (2.6) exists if and only if the following limit exists:

$$\lim_{\varepsilon\to 0}\int_{[\partial D\times(D'\setminus D_l)]\setminus\{|\zeta - z|<\varepsilon\}}\frac{1}{|\zeta - z|^{2n-1}}\,dv\,ds. \qquad (2.7)$$

Now (2.7) is a proper integral; hence we may go over to the iterated integal, and the existence of the limit (2.7) is equivalent to that of

$$\lim_{\varepsilon\to 0}\int_{\partial D} g_\varepsilon(\zeta)\,ds,\quad \text{where } g_\varepsilon(\zeta) = \int_{(D'\setminus D_l)\setminus\{|\zeta - z|<\varepsilon\}}\frac{1}{|\zeta - z|^{2n-1}}\,dv. \qquad (2.8)$$

But $\lim_{\varepsilon\to 0} g(\zeta)$ exists, and

$$0\leqslant g_\varepsilon(\zeta)\leqslant\int_{D'\setminus D_l}\frac{dv}{|\zeta - z|^{2n-1}}\leqslant\int_V\frac{dv}{|z|^{2n-1}}<\infty,$$

where $V = \bigcup_{\zeta \in \partial D}[(D^l \setminus D_l) - \zeta] \Subset C^n$; hence the limit (2.8) exists by the Lebesgue dominated convergence theorem, and so (2.6) exists too.

Since the integrand in (2.5) is majorized by that in (2.6), we may reverse the order of integration in (2.5), and the function

$$\int_{\partial D_\zeta} \frac{f(\zeta, z)(\bar{\zeta}_k - \bar{z}_k)}{|\zeta - z|^{2n}} d\bar{\zeta}\,[j] \wedge d\zeta$$

is integrable; hence the limit in (2.5) equals

$$\lim_{l \to \infty} \int_{(D^l \setminus D_l)_z} d\bar{z} \wedge dz \int_{\partial D_\zeta} \frac{f(\zeta, z)(\zeta_k - z_k)}{|\zeta - z|^{2n}} d\bar{\zeta}\,[j] \wedge d\zeta = 0,$$

since the measure of $D^l \setminus D_l$ tends to zero as $l \to \infty$.

4°. In this subsection, we shall obtain an analogue of Proposition 2.2 for forms of type $(p, 0)$. Let D be a domain with smooth boundary, $z^0 \in \partial D$, and $\gamma \in C_{(p,0)}(\partial D)$. In the interior of the right circular cone V_{z^0} with vertex z^0, the normal to ∂D at z^0 as axis, and angle between the axis and generator $\beta < \pi/2$, we choose two points $z \in D$ and $z' \notin \bar{D}$ such that $a|z - z^0| \leqslant |z' - z^0| \leqslant b|z - z^0|$, where a and b are constants with $0 < a \leqslant b < \infty$. Consider the limit

$$\lim_{z, z' \to z^0} [\gamma^+(z) - \gamma^-(z')]. \tag{2.9}$$

THEOREM 2.6. *The limit* (2.9) *exists and is equal to* $\gamma(z^0)$. *The convergence is uniform in* $z^0 \in \partial D$ *if* β, a *and* b *are fixed.*

Note that for $n = 1$ this is a strengthening of Proposition 2.2.

PROOF. Exactly as in the proof of (1.1) we can show that it is enough to prove the theorem for (0, 0)-forms, i.e. functions. Thus let $\gamma \in C(\partial D)$. Then

$$\gamma^+(z) - \gamma^-(z') = I^1_{0,0}(D, \gamma(\zeta) - \gamma(z^0))(z) - I^1_{0,0}(D, \gamma(\zeta) - \gamma(z^0))(z') \\ + \gamma(z^0)[I^1_{0,0}(D, 1)(z) - I^1_{0,0}(D, 1)(z')].$$

By (1.1), $I^1_{0,0}(D, 1)(z) = 1$ and $I^1_{0,0}(D, 1)(z') = 0$; hence we need only prove that

$$\lim_{z, z' \to z^0} \int_{\partial D} (\gamma(z) - \gamma(z^0))(U_{0,0}(\zeta, z) - U_{0,0}(\zeta, z')) = 0. \tag{2.10}$$

By a unitary transformation and a translation, we make take z^0 to 0, and the tangent plane to ∂D at z^0 to the plane $\alpha = \{\xi \in \mathbf{C}^n \colon \operatorname{Im} \xi_n = 0\}$. Then the left side of (2.10) is unchanged, and ∂D will be defined near 0 by the system of equations $'\zeta = {}'w$ and $\zeta_n = u_n + i\rho(w)$. Here $'w = (w_1, \ldots, w_{n-1})$, $w = ('w, u_n) \in \alpha$, and $\rho \in C^1(U)$, where U is a neighborhood of 0 in the plane α

and $\rho(w) = o(|w|)$. Let $\tilde{z}$ and $\tilde{z}'$ denote the projections of z and z' on the Im ξ_n-axis. Then

$$\begin{gathered} |z - \tilde{z}| < \tan\beta|\tilde{z}|, \qquad |z' - \tilde{z}'| < \tan\beta|\tilde{z}'|, \\ |z| < |\tilde{z}|/\cos\beta, \qquad |z'| < |\tilde{z}'|/\cos\beta, \\ a|\tilde{z}|/\cos\beta \leqslant |\tilde{z}'| \leqslant b\cos\beta|\tilde{z}|. \end{gathered} \tag{2.11}$$

We fix an $\varepsilon > 0$, and choose a $(2n-1)$-ball B in α such that

1) $B \subset U$;

2) $|\gamma(\zeta(w)) - \gamma(0)| < \varepsilon$, $w \in U$;

3) $|w - \tilde{z}| \leqslant C|\zeta(w) - z|$ and $|w - \tilde{z}'| \leqslant C|\zeta(w) - z|$, for all $w \in U$ and a suitable constant C. Condition 2) is satisfied because γ is continuous, and condition 3) because of the relations

$$|\zeta(w) - w| = |\rho(w)| = \rho(|w|), \qquad |w| < |w - \tilde{z}|$$

and the similar relations with z' in place of z. Observe that C and the radius of B can be chosen to be independent of $z_0 \in \partial D$.

Let $B' = \{\zeta \in \partial D : \zeta = \zeta(w), w \in B\}$.

Let us estimate the integral over B'. Using (1.4), we get

$$U_{0,0}(\zeta, z) - U_{0,0}(\zeta, z') = \frac{(n-1)!}{(2\pi i)^n} \sum_{k=1}^{n} (-1)^{k-1} \times \left(\frac{\bar{\zeta}_k - \bar{z}_k}{|\zeta - z|^{2n}} - \frac{\bar{\zeta}_k - \bar{z}'_k}{|\zeta - z'|^{2n}} \right) d\bar{\zeta}[k] \wedge d\zeta.$$

Now we use condition 3) and the inequality $|\zeta(w)| < C_1|w| < C_1|w - \tilde{z}|$ to obtain

$$\begin{aligned} \left| \frac{\bar{\zeta}_k}{|\zeta - z|^{2n}} - \frac{\bar{\zeta}_k}{|\zeta - z'|^{2n}} \right| &= \frac{|\zeta - z| - |\zeta - z'|}{|\zeta - z||\zeta - z'|} \sum_{j=0}^{2n-1} \frac{|\zeta_k|}{|\zeta - z|^j |\zeta - z'|^{2n-1-j}} \\ &\leqslant C_1 C^{2n+2} \sum_{j=0}^{2n-1} \frac{|z| + |z'|}{|w - \tilde{z}|^j |w - \tilde{z}'|^{2n-j}}. \end{aligned} \tag{2.12}$$

We may assume that $a_1 = a(\cos\beta)^{-1} < 1$; then (2.11) implies $|w - \tilde{z}| \geqslant |w - a_1\tilde{z}|$ and $|w - \tilde{z}'| \geqslant |w - a_1\tilde{z}|$. Thus from (2.12) we get

$$\left| \frac{\bar{\zeta}_k}{|\zeta - z|^{2n}} - \frac{\bar{\zeta}'_k}{|\zeta - z'|^{2n}} \right| \leqslant d\frac{|\tilde{z}|}{|w - a_1\tilde{z}|^{2n}}, \tag{2.13}$$

where d depends only on a, b, C, C_1 and β.

Similarly,

$$\left|\frac{\bar{z}_k}{|\zeta - z|^{2n}} - \frac{\bar{z}'_k}{|\zeta - z'|^{2n}}\right| \leqslant \frac{|z_k|}{|\zeta - z|^{2n}} + \frac{|z'_k|}{|\zeta - z'|^{2n}} \leqslant \frac{d_1|\tilde{z}|}{|w - a_1\tilde{z}|^{2n}}, \tag{2.14}$$

where d_1 depends only on a, b, C, C_1 and β.

Finally,

$$d\sigma \leqslant d_2 ds. \tag{2.15}$$

Here $d\sigma$ is the surface element on ∂D, and ds that on α; d_2 is a quantity not depending on $z^0 \in \partial D$. Now, from condition 2) and (2.13)–(2.15) we get

$$\left|\int_{B'_\zeta} (\gamma(\zeta) - \gamma(0))(U_{0,0}(\zeta, z) - U_{0,0}(\zeta, z'))\right|$$

$$\leqslant \frac{(n-1)!}{2\pi^n} d_2(d + d_1)\varepsilon \int_{B_w} \frac{|\tilde{z}|\,ds}{|w - a_1\tilde{z}|^{2n}} \leqslant d_3\varepsilon \int_{\mathbf{R}_w^{2n-1}} \frac{|\tilde{z}|\,ds}{|w - a_1\tilde{z}|^{2n}}.$$

Here $w = (w_1, \ldots, w_{n-1}, u_n)$ and $\tilde{z} = (0, \ldots, 0, iy_n)$; hence $|\tilde{z}| = |y_n|$ and

$$|w - a_1\tilde{z}|^{2n} = \left(|w_1|^2 + \cdots + |w_{n-1}|^2 + u_n^2 + a_1^2 y_n^2\right)^n.$$

Thus the last integral differs from the Poisson integral only by a constant factor and is independent of $\tilde{z}$.

To finish the proof, it remains only to note that

$$\lim_{z, z' \to z^0} \int_{(\partial D \setminus B')_\zeta} (f(\zeta) - f(0))(U_{0,0}(\zeta, z) - U_{0,0}(\zeta, z')) = 0,$$

and that this limit is approached uniformly since the radius of the ball B does not depend on $z_0 \in \partial D$.

COROLLARY 2.7. *If $\partial D \in C^{1,\lambda}$ and $\gamma \in C^{0,\lambda}_{(p,0)}(\partial D)$, then $\gamma = \gamma^+|_{\partial D} - \gamma^-|_{\partial D}$ on ∂D.*

This follows immediately from Theorem 2.6 and Corollary 2.4.

We note that, if $\gamma \in C(\partial D)$, then $\gamma^\pm$ are in general not extendable to $\bar{D}$ as continuous functions, and it is not possible to obtain an assertion analogous to Corollary 2.7 for continuous forms. This is shown by the following example.

EXAMPLE 2.8.([4]) Let D be a domain such that ∂D contains a $(2n-1)$-dimensional ball B lying in the plane $\{y_n = 0\}$, with radius $R < 1$ and center 0. On B we set

$$\gamma(\zeta) = \frac{\zeta_n}{|\zeta|\,|\ln|\zeta||}.$$

([4]) This example has been constructed in analogy with the one given by Gyunter ([1], Chapter II, §7).

Then $\gamma \in C(B)$. We extend γ as a continuous function on ∂D. The behavior of the functions $\gamma^{\pm}$ near B is determined by the integral

$$\gamma_1(z) = \int_{B_\zeta} \gamma(\zeta) U_{0,0}(\zeta, z).$$

Let us show that γ_1 is not bounded in any neighborhood of 0. Indeed, let $z = (0, \ldots, 0, iy_n)$. Then

$$\gamma_1(z) = C \int_{B_\zeta} \frac{\zeta_n(\bar{\zeta}_n - iy_n)\, d\bar{\zeta}\,[n] \wedge d\zeta}{|\zeta|\,|\ln|\zeta||\,|\zeta - z|^{2n}}, \qquad C \neq 0.$$

The integrals of the other terms vanish since $d\bar{\zeta}_n \wedge d\zeta_n = 0$ on B. Furthermore,

$$\int_{B_\zeta} \frac{\zeta_n y_n\, d\bar{\zeta}\,[n] \wedge d\zeta}{|\zeta|\,|\ln|\zeta||\,|\zeta - z|^{2n}} = 0,$$

since the integrand is odd and the domain of integration is symmetric about 0. Transforming to spherical coordinates, we have

$$\gamma_1(z) = C_1 \int_S \cos^2(\tau x_n)^\wedge\, ds \int_0^R \frac{r^{2n-1} dr}{|\ln r| \cdot |r^2 + y_n^2|^n}. \tag{2.16}$$

Here, S is the $(2n-2)$-sphere of radius 1, ds its surface element, $(\tau x_n)^\wedge$ the angle between the x_n-axis and the radius vector of τ, and $C_1 \neq 0$ a constant.

The first integral on the right in (2.16) is positive; hence

$$|\gamma_2(z)| \geqslant C_2 \int_0^R \frac{r^{2n-1} dr}{|\ln r|\,|r^2 + y_n^2|^n} \geqslant C_2 \int_\delta^R \frac{r^{2n-1} dr}{|\ln r|\,|r^2 + y_n^2|^n}, \qquad C_2 > 0.$$

As $y_n \to 0$, this integral converges to

$$\int_\delta^R \frac{r^{2n-1} dr}{|\ln r|\, r^{2n}} = \ln\left|\frac{\ln \delta}{\ln R}\right|,$$

so that, for sufficiently small y_n,

$$|\gamma_1(z)| \geqslant \frac{1}{2} C_2 \ln\left|\frac{\ln \delta}{\ln R}\right|. \tag{2.17}$$

Since the right side of (2.17) tends to infinity as $\delta \to 0$, it follows that γ_1 is unbounded in every neighborhood of 0.

5°. For $p < n$ and $q > 0$, the assertions analogous to those of Propositions 2.1 and 2.2 and Theorems 2.5 and 2.6 are false, as the following example shows.

EXAMPLE 2.9. Let D be the same domain as in Example 2.8, $\alpha = d\bar{z}_{n-q+1} \wedge \cdots \wedge d\bar{z}_{n-1} \wedge dz_1 \wedge \cdots \wedge dz_p$ and $\gamma = d\bar{z}_n \wedge \alpha$. Then $\gamma \in C^\infty_{(p,q)}(\partial D)$. For

$z \in B$,

$$\gamma^+|_{\partial D} - \gamma^-|_{\partial D} = \gamma_1^+|_{\partial D} - \gamma_1^-|_{\partial D},$$

where

$$\gamma_1^{\pm} = \int_{B_\zeta} \gamma(\zeta) \wedge U_{p,q}(\zeta, z).$$

Since $\alpha(\zeta) \wedge U_{p,q}$ is of type $(n, n-2)$ in ζ, we have

$$\gamma(\zeta) \wedge U_{p,q}(\zeta, z)\,|_B = 0,$$

i.e. $\gamma_1^{\pm} \equiv 0$; hence for $z \in B$ we have $\gamma^+|_{\partial D} - \gamma^-|_{\partial D} = 0$, but $\gamma|_{\partial D} \neq 0$.

For such forms, the jump theorem can be obtained in the following way:

Let D be a domain with a piecewise smooth boundary, and $\gamma \in C^1_{(p,q)}(\partial D)$. Let $\tilde{\gamma}$ denote any form in $C^1_{(p,q)}(\overline{D})$ such that $\tilde{\gamma}|_{\partial D} = \gamma$, and let

$$\gamma_{\pm}(z^{\pm}) = I^1_{p,q}(D, \gamma)(z^{\pm}) - \bar{\partial} I^2_{p,q-1}(D, \tilde{\gamma})(z^{\pm}),$$
$$z^+ \in D, \qquad z^- \in C\overline{D}.$$

THEOREM 2.10. *$\gamma_{\pm}$ extend continuously to ∂D, and*

$$\gamma = \gamma_+|_{\partial D} - \gamma_-|_{\partial D}. \tag{2.18}$$

PROOF. By (1.1), we have for $z \in D$

$$\gamma_+(z) = \tilde{\gamma}(z) + I^2_{p,q}(D, \bar{\partial}\tilde{\gamma})(z) \tag{2.19}$$

and for $z \notin \overline{D}$

$$\gamma_-(z) = I^2_{p,q}(D, \bar{\partial}\tilde{\gamma})(z). \tag{2.20}$$

The integral that occurs in the definition of $I^2_{p,q}$ is absolutely convergent for all $z \in \mathbf{C}^n$; hence $I^2_{p,q}(D, \bar{\partial}\tilde{\gamma}) \in C_{(p,q)}(\mathbf{C}^n)$. Hence the forms $\gamma_{\pm}$ can be extended continuously to ∂D preserving (2.19) and (2.10). Substracting (2.20) from (2.19), we get (2.18).

6°. We now proceed to the additive problem of Riemann for forms. We formulate it as follows:

Can every $\gamma \in C^1_{(p,q)}(\partial D)$ be represented in the form

$$\gamma = \gamma_1|_{\partial D} - \gamma_2|_{\partial D}, \tag{2.21}$$

with $\gamma_1 \in Z^1_{(p,q)}(\overline{D})$ and $\gamma_2 \in Z^1_{(p,q)}(CD)$, and it not, what are the conditions on γ of the validity of such a decomposition?

THEOREM 2.11. I. *If $\partial D \in C^1$ and $\gamma \in C^1_{(p,q)}(\partial D)$, then the decomposition (2.21) holds only if there exists $\tilde{\gamma} \in C^1_{(p,q)}(\mathbf{C}^n)$ such that $\tilde{\gamma}|_{\partial D} = \gamma$ and $\bar{\partial}\tilde{\gamma} = 0$ on ∂D.*

II. *If* $\partial D \in C^{m+1,\lambda}$ *and* $\gamma \in C^{m,\lambda}(\partial D)$, $m \geqslant 1$, *then a sufficient condition for* (2.21) *to hold is the existence of a* $\tilde{\gamma}_1 \in C^1_{p,q}(C^n)$ *such that* $\tilde{\gamma}_1|_{\partial D} = \gamma$ *and* $\bar{\partial}\tilde{\gamma}_1 \wedge dz_J|_{\partial D} = 0$ *for any* $J = (j_{p+1},\ldots,j_n)$. *The forms* γ_1 *and* γ_2 *can then be chosen in* $C^{m,\lambda'}_{(p,q)}$, $0 < \lambda' < \lambda$, *with* $\gamma_2 = o(|z|^{1-2n})$ *as* $|z| \to \infty$.

PROOF. I. We first extend γ_1 and γ_2 to all of $\mathbf{C}^n$ as forms of class $C^1_{(p,q)}(\mathbf{C}^n)$. We continue to denote these extensions by γ_1 and γ_2, and set $\tilde{\gamma} = \gamma_1 - \gamma_2$. Then $\tilde{\gamma}|_{\partial D} = \gamma$, and on ∂D

$$\bar{\partial}\tilde{\gamma} = \bar{\partial}\gamma_1 - \bar{\partial}\gamma_2 = 0,$$

since $\bar{\partial}\gamma_1 = 0$ for $z \in \bar{D}$ and $\partial\gamma_2 = 0$ for $z \notin D$.

II. We shall show that the forms $\gamma_\pm$ defined in subsection 5° are $\bar{\partial}$-closed. Indeed,

$$\bar{\partial}\gamma_\pm = \bar{\partial}I^1_{p,q}(D,\gamma) = \bar{\partial}_z\int_{\partial D_\zeta}\gamma(\zeta) \wedge U_{p,q}(\zeta,z) = \int_{\partial D_\zeta}\gamma(\zeta) \wedge \bar{\partial}_z U_{p,q}(\zeta,z).$$

Since $\gamma \wedge U_{p,q}$ is of type $(n, n-1)$ in ζ, it follows by Lemma 1.2 and Stokes' formula that

$$\begin{aligned}\bar{\partial}\gamma_\pm &= (-1)^q\int_{\partial D_\zeta}\gamma(\zeta) \wedge \bar{\partial}_\zeta U_{p,q+1}(\zeta,z)\\ &= (-1)^q\int_{\partial D_\zeta}\bar{\partial}\tilde{\gamma}_1(\zeta) \wedge U_{p,q+1}(\zeta,z) = 0.\end{aligned}$$

For sufficiently large z, the coefficients of $U_{p,q}(\zeta,z)$ and $U_{p,q-1}(\zeta,z)$ are majorized uniformly in $\zeta \in \bar{D}$ by a function of the form $C|z|^{1-2n}$. An application of Corollary 2.4 and Theorem 2.10 concludes the proof.

For $q = n-1$, we can get more information on the forms γ_1 and γ_2 occurring in (2.21). Let us denote by $B_{p,n-1}(D)$ the set of $\bar{\partial}$-closed forms of the type

$$\alpha = \sum_J \sum_{k=1}^n (-1)^{k-1}\frac{\partial G_J}{\partial z_k} d\bar{z}[k] \wedge dz_J,$$

where the $J = (j_1,\ldots,j_p)$ are multi-indices and the G_J are harmonic functions in D.

THEOREM 2.12. *If* $\partial D \in C^{m+1,\lambda}$ *and* $\gamma \in C^{m,\lambda}_{(p,n-1)}(\partial D)$, $m \geqslant 1$, *then* (2.21) *holds with* $\gamma_1 \in Z^{m,\lambda'}_{(p,n-1)}(\bar{D})$ *and* $\gamma_2 \in B_{p,n-1}(C\bar{D}) \cap C^{m,\lambda'}_{(p,n-1)}(CD)$, $0 < \lambda' < \lambda$, $\gamma_2 = O(|z|^{1-2n})$ *as* $|z| \to \infty$.

If $p = n$, *then* $\gamma_1 \in B_{n,n-1}(D)$.

PROOF. Let $\tilde{\gamma} \in C^{m,\lambda}_{(p,n-1)}(\bar{D})$ and $\tilde{\gamma}|_{\partial D} = \gamma$. Then $I^2_{p,n-2}(D, \tilde{\gamma}) \in C^{m+1,\lambda'}_{(p,n-2)}(CD)$ (see Corollary 2.4), i.e. there exists $\beta \in C^{m+1,\lambda'}_{(p,n-2)}(\mathbf{C}^n)$ such that $\beta = I^2_{p,n-2}(D, \tilde{\gamma})$ on CD. We set $\gamma_1 = \gamma_+ + \bar{\partial}\beta$ and $\gamma_2 = I^1_{p,n-1}(D, \gamma)$ ($\gamma_\pm$ have been defined in 5°). We have (see (1.4))

$$U_{p,n-1}(\zeta, z) = \frac{(n-1)!}{(2\pi i)^n} \sum_J \sum_{k=1}^n (-1)^{k-1} \times \frac{\partial}{\partial z_k} \frac{\sigma(J)\, d\zeta[J]}{(n-1)\,|\zeta - z|^{2n-2}} d\bar{z}[k] \wedge dz;$$

hence $I^1_{p,n-1}(D, \gamma) \in B_{p,n-1}(CD)$. By Theorem 2.9,

$$\begin{aligned} \gamma_1|_{\partial D} - \gamma_2|_{\partial D} &= \gamma_+|_{\partial D} - \left(I^1_{p,n-1}(D, \gamma) - \bar{\partial} I^2_{p,n-2}(D, \tilde{\gamma})\right) \\ &= \gamma_+|_{\partial D} - \gamma_-|_{\partial D} = \gamma. \end{aligned}$$

The remaining properties of γ_1 and γ_2 are proved in the same way as in Theorem 2.11.

For proving the second part of the theorem, we have to use Theorem 2.5.

In connection with Theorem 2.12, we note

COROLLARY 2.13. *Let* $\partial D \in C^{m+1,\lambda}$ *and* $\gamma \in Z^{m,\lambda}_{(p,n-1)}(\bar{D})$, $m \geqslant 1$; *then* $\gamma = \gamma_1 + \bar{\partial}\beta$, *where* $\gamma_1 \in B_{(p,n-1)}(D) \cap Z^{m,\lambda'}_{(p,n-1)}(\bar{D})$ *and* $\beta \in C^{m+1,\lambda'}_{(p,n-2)}(\bar{D})$, $0 < \lambda' < \lambda$.

This follows easily from (1.1) and Corollary 2.4.

§3. Characterization of the trace of a holomorphic form on the boundary of a domain

1°. Proposition 2.2 allows us to give a necessary and sufficient condition for a continuous function defined on D to be extendable holomorphically into D (see Muskhelishvili [1], §35). Theorem 2.6, being an analogue of Proposition 2.2, can be similarly used for a similar purpose in the case of functions (or more generally forms) of several complex variables.

THEOREM 3.1. *Let* D *be a domain in* $\mathbf{C}^n$ *with smooth boundary, and let* $\gamma \in C_{(p,0)}(\partial D)$. *Then a form* $\tilde{\gamma} \in AC_p(D)$ *such that* $\tilde{\gamma}|_{\partial D} = \gamma$ *exists if and only if*

$$\int_{\partial D} \gamma \wedge \theta = 0 \tag{3.1}$$

for any $\theta \in Z^{\infty}_{(n-p,n-1)}(\bar{D})$.

PROOF. *Necessity.* Let D_l be the sequence of domains defined in the proof of Theorem 2.5. Then

$$\int_{\partial D} \gamma \wedge \theta = \lim_{l\to\infty} \int_{\partial D_l} \tilde{\gamma} \wedge \theta = \lim_{l\to\infty} \int_{D_l} d(\tilde{\gamma} \wedge \theta) = 0$$

for any $\theta \in Z^{\infty}_{(n-p,n-1)}(\overline{D})$, since

$$d(\tilde{\gamma} \wedge \theta) = \bar{\partial}(\tilde{\gamma} \wedge \theta) = \bar{\partial}\tilde{\gamma} \wedge \theta + (-1)^p \tilde{\gamma} \wedge \bar{\partial}\theta = 0.$$

Sufficiency. By Lemma 1.2, $\bar{\partial}_\zeta U_{p,0}(\zeta, z) = 0$ for $\zeta \neq z$; hence it follows from the definition of γ^- (§2, 2°) and (3.1) that $\gamma^- \equiv 0$ for $z \notin \overline{D}$. Then Theorem 2.6 yields

$$\lim_{\substack{z \to z^0 \in \partial D \\ z \in D}} \gamma^+(z) = \gamma(z^0),$$

the limit being approached uniformly in z_0 if z approaches z^0 along the norm to ∂D. Let us show that the form

$$\tilde{\gamma} = \begin{cases} \gamma, & z \in \partial D, \\ \gamma^+, & z \in D, \end{cases}$$

is continuous on $\overline{D}$. Indeed, $\gamma \in C_{n-p}(\partial D)$, and so for each $\varepsilon > 0$ there exists $\delta_1 > 0$ such that, for all $z^0, z' \in \partial D$ with $|z' - z^0| < \delta_1$,

$$|\gamma(z') - \gamma(z^0)| < \varepsilon/2. \text{ (}^5\text{)}$$

Choose δ, $0 < \delta < \delta_1/2$, such that if $z' \in \partial D$ and z lies on the inward normal to ∂D at z' and $|z - z'| < \delta$, then

$$|\gamma^+(z) - \gamma(z')| < \varepsilon/2.$$

Now let $z^0 \in \partial D$, $z \in D$, and $|z - z^0| < \delta$. Choose $z' \in \partial D$ such that $\rho(z, \partial D) = |z - z'|$. Then z lies on the normal to ∂D at z', and

$$|z' - z^0| \leqslant |z' - z| + |z - z^0| = \rho(z, \partial D) + |z - z^0| < 2\delta < \delta_1.$$

Then

$$|\tilde{\gamma}(z) - \gamma(z^0)| \leqslant |\tilde{\gamma}(z) - \gamma(z')| + |\gamma(z') - \gamma(z^0)| < \varepsilon.$$

Thus we have proved the continuity of $\tilde{\gamma}$ at the boundary points. The continuity at interior points is obvious.

([5]) By the absolute value of a form, we mean the maximum of the absolute values of its coefficients.

It remains to prove that $\tilde{\gamma} \in A_p(D)$. For a fixed $z \in D$, on account of Lemma 1.2 we have

$$\bar{\partial}\tilde{\gamma}(z) = \bar{\partial}\gamma + (z) = \int_{\partial D_\zeta} \gamma(\zeta) \wedge \bar{\partial}_z U_{p,0}(\zeta, z)$$

$$= (-1)^p \int_{\partial D_\zeta} \gamma(\zeta) \wedge \bar{\partial}_\zeta U_{p,1}(\zeta, z).$$

Let $U_z'(\zeta) \in C^\infty_{n-p,n-2}(\mathbf{C}^n)$ be such that $U_z'(\zeta) = U_{p,1}(\zeta, z)$ in some neighborhood of ∂D. Then $\bar{\partial} U_z'(\zeta) \in Z^\infty_{n-p,n-1}(\mathbf{C}^n)$, and (3.1) yields

$$\bar{\partial}\tilde{\gamma}(z) = (-1)^p \int_{\partial D_\zeta} \gamma(\zeta) \wedge \bar{\partial} U_z'(\zeta) = 0.$$

2°. Suppose $f \in C^1(\partial D)$ satisfies the tangential Cauchy-Riemann equations, i.e.

$$df \wedge dz\,|_{\partial D} = 0. \tag{3.2}$$

Then, for any $\chi \in C^\infty_{(n,n-2)}(\bar{D})$, by Stokes' theorem,

$$\int_{\partial D} f\bar{\partial}\chi = \int_{\partial D} df \wedge \chi = 0;$$

hence it is natural to call $\gamma \in C_{p,0}(\partial D)$ *a weak solution of the tangential Cauchy-Riemann equations* if

$$\int_{\partial D} \gamma \wedge \bar{\partial}\chi = 0 \quad \text{for all } \chi \in C^\infty_{(n-p,n-2)}(\bar{D}).$$

THEOREM 3.2. *Let D be a domain in $\mathbf{C}^n$, $n \geqslant 2$, with smooth connected boundary. Then the space of weak solutions of the tangential Cauchy-Riemann equations is identical with the space of traces on ∂D of forms in $AC_p(D)$.*

The proof is exactly the same as that of Theorem 3.1. We need only show that, if ∂D is connected and γ is a weak solution of the tangential Cauchy-Riemann equations, then $\gamma^- \equiv 0$ in $C\bar{D}$.

Let B be a ball with $D \Subset B$. Since B is a domain of holomorphy, and $\bar{\partial}_\zeta U_{p,0}(\zeta, z) = 0$ for $\zeta \in B$ and $z \notin B$, we have $U_{p,0}(\zeta, z) = \bar{\partial}_\zeta \chi_z(\zeta)$ and $\chi_z \in C^\infty_{(n-p,n-2)}(B)$, so that for $z \notin B$

$$\gamma^-(z) = \int_{\partial D_\zeta} \gamma(\zeta) \wedge U_{p,0}(\zeta, z) = \int_{\partial D_\zeta} \gamma(\zeta) \wedge \bar{\partial}_\zeta \chi_z(\zeta) = 0. \tag{3.3}$$

∂D being smooth and connected, $C\bar{D}$ is connected; and γ^- is harmonic since the coefficients of $U_{p,0}(\zeta, z)$ are. Hence (3.3) and the uniqueness theorem imply that $\gamma^- \equiv 0$ in $C\bar{D}$.

§4. Some cases of solvability of the $\bar\partial$-problem

1°. As we know, one of the central questions in the theory of functions of several complex variables is the solution of the $\bar\partial$-problem, i.e. of the equation

$$\bar\partial\alpha = \gamma, \tag{4.1}$$

where γ is a (p, q)-form, $q \geqslant 1$. Many questions in complex analysis ar either closely related to this problem, or can be reduced to it. For example, D is a domain of holomorphy if and only if the $\bar\partial$-problem is solvable in D for any $\gamma \in Z^{\infty}_{(p,q)}(D)$, $1 \leqslant p \leqslant n$ (Hörmander [1], §4.2). We need the answers to the following questions.

1. Let $\gamma \in Z^k_{(p,q)}(\mathbf{C}^n)$ and $\operatorname{supp}\gamma = K$. Under what conditions on γ and K can (4.1) be solved with the same support?

2. Let $\gamma \in Z^k_{(p,q)}(\bar D)$. Under what conditions on D can (4.1) be solved with the same smoothness on $\bar D$?

2°. To answer these questions, we need the following result of Khenkin [1],([6]) sharpened by Øvrelid [1]:

Let D be a bounded strictly pseudoconvex domain with boundary of class C^m. Then there exist a domain of holomorphy V, $D \Subset V$, a vector-valued function $P(\zeta, z)$, and positive constants ε, C_1 and C_2 such that

1. $P(\zeta, z) \in C^{m-1}((V\setminus D) \times V)$, and
2. $P(\zeta, z)$ is holomorphic in $z \in V$, for each fixed $\zeta \in V\setminus D$.
3. $\Phi(\zeta, z) = \langle P(\zeta, z), \zeta - z\rangle \neq 0$ for $\zeta \in V\setminus D$, $z \in \bar D\setminus\{\zeta\}$.
4. $$|\Phi(\zeta, z)| \geqslant C_1(\rho(\zeta) - \rho(z)) + C_2|\zeta - z|^2$$

for $\zeta \in V\setminus D$, $z \in \bar D$, $|\zeta - z| < \varepsilon$.

In the rest of this section, we shall denote by u the vector-valued function

$$\frac{P}{\Phi} = \left(\frac{P_1}{\Phi}, \ldots, \frac{P_n}{\Phi}\right).$$

We recall that a domain $D \subset \mathbf{C}^n$ with boundary of class C^2 is said to be *strictly pseudoconvex* if $D = \{z \in \mathbf{C}^n\colon \rho(z) < 0\}$, where $\rho \in C^2(\mathbf{C}^n)$, $\operatorname{grad}\rho \neq 0$ on ∂D, and the Levi form

$$\sum_{i,j=1}^{n} \frac{\partial^2\rho}{\partial z_i \partial\bar z_j}(z) w_i \bar w_j$$

is positive for all $z \in \partial D$ and all $w \neq 0$ such that

$$\frac{\partial\rho}{\partial z_1}(z)\cdot w_1 + \cdots + \frac{\partial\rho}{\partial z_n}(z)\cdot w_n = 0.$$

([6]) We state the result only in the form we need.

3°. Let us consider integrals of the form

$$\int_{D_z} \gamma(z) \wedge D_{1,1,q,r,n-q-r-2}(t, u, \bar\partial_z t, \bar\partial_\zeta u, \bar\partial_\zeta t) \wedge D_{p,n-p}(\partial z, \partial\zeta),$$

$$\int_{D_\zeta} \gamma(\zeta) \wedge D_{1,1,q,r,n-q-r-2}(t, u, \bar\partial_z t, \bar\partial_\zeta u, \bar\partial_\zeta t) \wedge D_{p,n-p}(\partial z, \partial\zeta),$$

where $t = (\bar\zeta - \bar z)/|\zeta - z|^2$ as usual, and γ is a form of the appropriate dimension.

Our interest is in the following question: is it possible, by imposing suitable conditions on γ, to extend the forms defined by these integrals to forms on V, differentiable as many times as we need?

Let us first explicitly write down the singularity of the integrals:

LEMMA 4.1.

$$D_{1,1,q,r,n-q-r-2}(u, t, \bar\partial_z t, \bar\partial_\zeta u, \bar\partial_\zeta t) = \sum_{k=1}^{n} \frac{(\bar\zeta_k - \bar z_k)\mu_k^{q,r}(\zeta, z)}{|(\zeta - z)|^{2(n-r-1)}\Phi^{r+1}},$$

where the $\mu_k^{q,r}$ are forms of type $(0, q)$ in z and $(0, n-q-2)$ in ζ, with coefficients in $C^{m-2}((V\setminus D) \times V)$.

The proof proceeds exactly like that of (1.4).

We can now answer the question asked above.

LEMMA 4.2. *Let $m \geqslant 3$ and $\gamma \in C^{m-3,\lambda}_{p,q}(\bar V)$.*

1. *If $\gamma \equiv 0$ on $V\setminus D$, then*

$$\int_{D_z} \gamma(z) \wedge D_{1,1,n-q,r-2}(t, u, \bar\partial_z t, \bar\partial_\zeta u, \bar\partial_\zeta t) \wedge D_{n-p,p}(\partial z, \partial\zeta) \in C^{m-2}_{(p,q-2)}(V\setminus D).$$

2. *If $\gamma \equiv 0$ on D, then*

$$\int_{(V\setminus D)_\zeta} \gamma(\zeta) \wedge D_{1,1,q-r,r,n-q-2}(t, u, \bar\partial_z t, \bar\partial_\zeta u, \bar\partial_\zeta t) \wedge D_{p,n-p}(\partial z, \partial\zeta) \in C^{m-2}_{(p,q-2)}(\bar D).$$

PROOF. We shall prove the first part; the proof of the second part is entirely similar. It is sufficient to prove the lemma for $m < \infty$.

From Lemma 4.1, we see that our assertion is a consequence of the following:

If $h \in C^{m-3,\lambda}(\bar V)$, $h \equiv 0$ on $V\setminus D$, and $\varphi \in C^{m-2}((\bar V\setminus D) \times \bar V)$, then

$$I(\zeta) = \int_{D_z} \frac{h(z)\varphi(\zeta, z)(\bar\zeta_k - \bar z_k)}{\Phi^{r+1}|\zeta - z|^{2(n-r-1)}} d\bar z \wedge dz \in C^{m-2}(V\setminus D). \tag{4.2}$$

For $\zeta \in V\setminus\bar{D}$, (4.2) is a proper integral, so we may differentiate under the integral sign $m-2$ times. Hence $I(\zeta) \in C^{m-2}(V\setminus\bar{D})$.

The derivatives of the integrand are linear combinations of expressions of the form

$$\psi = h(z) D_\zeta^{I,I'} \frac{1}{\Phi^{r+1}} \cdot D_\zeta^{J,J'} \frac{\bar\zeta_k - \bar z_k}{|\zeta - z|^{2(n-r-1)}} \chi(\zeta, \bar z),$$

where $\|I + I' + J + J'\| \leqslant m-2$ and $\chi(\zeta, z)$ is continuous on $(\bar V\setminus D) \times \bar V$. And we have

$$\left| D_\zeta^{I,I'} \frac{1}{(\Phi(\zeta, z))^{r+1}} \right| \leqslant \frac{M_1}{|\Phi(\zeta, z)|^{\|I+I'\|+r+1}},$$

$$\left| D_\zeta^{J,J'} \frac{\bar\zeta_k - \bar z_k}{|\zeta - z|^{2(n-r-1)}} \right| \leqslant \frac{M_1}{|\zeta - z|^{2(n-r-1)+\|J+J'\|-1}},$$

where M_1 does not depend on $(\zeta, z) \in (\bar V\setminus D) \times \bar D$.

We have $h = 0$ on $\partial D = \{z\colon\ \rho(z) = 0\}$; hence by Hadamard's lemma (Arnol'd [1], §2.6.4) there exists $h_1 \in C^{m-4,\lambda}(\bar V)$ such that $h = \rho h_1$; in turn, we have $h_1 = h\rho^{-1} \equiv 0$ on $V\setminus\bar D$, and so, by continuity, $h_1 \equiv 0$ on ∂D. Repeating this procedure $m-3$ times, we get $h = \rho^{m-3} h_2$, where $h_2 \in C^{0,\lambda}(\bar V)$ and $h_2 \equiv 0$ on $V\setminus D$. For λ' such that $0 < \lambda' < \lambda$, the function $h_0 = h_2 \cdot |\zeta - z|^{-\lambda'}$ lies in $C((\bar V\setminus D) \times \bar D)$; hence

$$|\psi| \leqslant \frac{M_2 |\rho(z)|^{m-3+\lambda'}}{|\Phi(\zeta, z)|^{\|I+I'\|+r+1} |\zeta - z|^{2(n-r-1)+\|J+J'\|-1}}$$

$$\leqslant \left| \frac{\rho(z)}{\Phi(\zeta, z)} \right|^{\|I+I'\|} \cdot \left| \frac{\rho(z)}{|\zeta - z|} \right| \|J + J'\| - 1 \cdot \frac{M_3}{|\Phi(\zeta, z)|^{r+1} \cdot |\zeta - z|^{2n-\lambda'-2(r+1)}},$$

where M_2 and M_3 are suitable constants.

Now, from property 4) of $P(\zeta, z)$ and the fact that $\rho(z) \leqslant 0$ for $z \in \bar D$ and $\rho(\zeta) \geqslant 0$ for $\zeta \in V\setminus D$, we have

$$|\rho(z)| \leqslant \frac{1}{C_1} |\Phi(\zeta, z)|, \qquad |\zeta - z|^2 \leqslant \frac{1}{C_2} |\Phi(\zeta, z)|$$

and, further, $|\rho(z)| \leqslant C_3 |\zeta - z|$ for $z \in \bar D$ and $\zeta \in V\setminus D$. Hence

$$|\psi| \leqslant \frac{M}{|\zeta - z|^{2n-\lambda'}},$$

where M is a constant. But this means that, for $\zeta \in \bar V\setminus D$, the integral of any derivative of the integrand in (4.2) is convergent uniformly in ζ, i.e. all derivatives of $I(\zeta)$ of order up to $m-2$ extend continuously to $\bar{V}\setminus D$. Now (4.2) is a simple consequence of the following assertion.

Let D be a bounded domain in $\mathbf{R}^n$ with boundary of class C^m, $m \geqslant 0$, and let $f \in C^m(D)$. Suppose f and all its derivatives of order up to m extend continuously to $\bar{D}$. Then $f \in C^m(\bar{D})$.

For $n = 2$, the proof of this statement is given in Fikhtengol'ts ([1], Paragraph 260). For $n > 2$ the proof is entirely analogous.

4°. We can now proceed to the solution of the problems formulated in 1°.

THEOREM 4.3. *Let D be a bounded strictly pseudoconvex domain with boundary of class C^{m+3}, and let $\gamma \in Z^{m,\lambda}_{(p,q)}(\mathbf{C}^n)$, $m \geqslant 0$. Suppose* $\operatorname{supp}\gamma \subset \bar{D}$. *Then*:

I. *For $q < n$, there exists $\alpha \in C^m_{(p,q-1)}(\mathbf{C}^n)$ such that* $\operatorname{supp}\alpha \subset \bar{D}$ *and* $\bar\partial\alpha = \gamma$.

II. *For $q = n$, an $\alpha \in C^m_{(p,n-1)}(\mathbf{C}^n)$ such that* $\operatorname{supp}\alpha \subset \bar{D}$ *and $\bar\partial\alpha = \gamma$ exists if and only if*

$$\int_D \mu \wedge \gamma = 0, \qquad \mu \in A_{n-p}(\bar{D}). \tag{4.3}$$

III. *If $\alpha \in C^{m,\lambda}_{(p,q-1)}(\mathbf{C}^n)$ and $\bar\partial\alpha = \gamma$*,(7) *then there exists $\beta \in C^{m+1}_{(p,q-l)}(\mathbf{C}^n)$ such that $\bar\partial\beta = \alpha$ on CD, $q \geqslant 2$.*

PROOF. The necessity of (4.3) is a simple consequence of Stokes' formula. Indeed, for any $\mu \in A_{n-p}(\bar{D})$.

$$\int_D \mu \wedge \gamma = \int_D d(\mu \wedge \alpha) = \int_{\partial D} \mu \wedge \alpha = 0.$$

Let us prove I and the remaining part of II. Let V be the domain of holomorphy introduced in 2°. Since $\operatorname{supp}\gamma \subset \bar{D}$ and $\bar\partial\gamma = 0$, for $z \in V$ we have by (1.1)

$$\gamma(z) = \bar\partial I^2_{p,q-1}(V,\gamma) = \bar\partial I^2_{p,q-1}(D,\gamma). \tag{4.4}$$

If $q = 1$, then $I^2_{p,q-1}(D,\gamma) \in A_p(C\bar{D})$, since $\operatorname{supp}\gamma \subset \bar{D}$. By Hartogs' theorem, $I^2_{p,q-1}(D,\gamma)$ can be extended to all of $\mathbf{C}^n$ as a form $\alpha_1 \in A_p(C^n)$. It now follows from (4.4) and Corollary 2.4 that $\alpha = I^2_{p,q-1}(D,\gamma) - \alpha_1$ is a form with the desired properties.

We now consider the case $q \geqslant 1$. Then, by Lemmas 1.3 and 1.4, for $z \in V\setminus\bar{D}$ we have

$$\begin{aligned} I^2_{p,q-1}(D,\gamma)(\zeta) &= \int_{D_z} \gamma(z) \wedge \bar\partial_z \mu^2_{q-1} \wedge \frac{1}{p!\,(n-p)!} D_{p,n-p}(\partial z, \partial\zeta) \\ &\quad + \bar\partial \int_{D_z} \gamma(z) \wedge \mu^1_{q-2} \wedge \frac{1}{p!(n-p)!} D_{p,n-p}(\partial z, \partial\zeta), \end{aligned} \tag{4.5}$$

(7) If $q = n$, it will be supposed that γ satisfies (4.3).

for $q < n$, and

$$I^2_{p,n-1}(D,\gamma)_1(\zeta) = \int_{D_z} \gamma(z) \wedge W_{p,0}(u,\zeta,z)$$

$$+\bar\partial \int_{D_z} \gamma(z) \wedge \mu_0 \wedge \frac{1}{p!\,(n-p)!} D_{p,n-p}(\partial z, \partial\zeta) \quad (4.6)$$

(for convenience, we have deviated from our usual notation and interchanged the places of ζ and z).

Using Stokes' formula and the equality $\gamma\,|_{\partial D} = 0$, we see that the first integral in (4.5) vanishes. Since $U(\zeta, z)$, and consequently $W_{p,0}(u, \zeta, z)$, are holomorphic in z on $\bar{D}$, the first integral in (4.6) also vanishes by virtue of (4.3). By Lemmas 1.3 and 1.4 we see that the second integrals in (4.5) and (4.6) are linear combinations of the integrals considered in Lemma 4.2; hence

$$I^2_{p,q-1}(D,\gamma) = \bar\partial\beta \quad \text{on } V \setminus D \quad (4.7)$$

and $\beta \in C^{m+1}_{(p,q-2)}(V)$.

Corollary 2.4 and (4.4) now show that $\alpha = I^2_{p,q-1}(D,\gamma) - \bar\partial\beta$ is a form with the desired properties.

It remains to prove III. By (1.1), we can represent α as

$$\alpha = I^1_{p,q-1}(V,\alpha) - I^2_{p,q-1}(V,\gamma) - \bar\partial I^2_{p,q-2}(V,\alpha).$$

As in the proof of Theorem 2.10, we can show that

$$I^1_{p,q-1}(V,\alpha) \in Z^\infty_{(p,q-1)}(V).$$

Since V is a domain of holomorphy,

$$I^1_{p,q-1}(V,\alpha) = \bar\partial\beta_1$$

with $\beta_1 \in C^\infty_{(p,q-2)}(V)$. The proof is completed by applying (4.7) and Corollary 2.4.

THEOREM 4.4. *Let D be a bounded strictly pseudoconvex domain with C^{m+2} boundary and $\gamma \in Z^{m,\lambda}_{(p,q)}(\bar{D})$, $m \geqslant 1$. Then there exists $\alpha \in C^m_{(p,q-1)}(\bar{D})$ such that $\bar\partial\alpha = \gamma$.*

PROOF. Consider the domain of holomorphy V introduced in 2°, and a $\tilde\gamma \in C^{m,\lambda}_{(p,q)}(\mathbf{C}^n)$ such that $\tilde\gamma \equiv \gamma$ on D and $\operatorname{supp}\tilde\gamma \subset V$. By (1.1),

$$\tilde\gamma = -I^2_{p,q}(V,\bar\partial\tilde\gamma) - \bar\partial I^2_{p,q-1}(V,\tilde\gamma).$$

For $q = n$ we have $I^2_{p,n}(V,\bar\partial\tilde\gamma) \equiv 0$; hence the theorem follows in this case from Corollary 2.4.

Now let $q \leqslant n-1$. Then, since $\bar\partial\tilde\gamma = 0$ on $\bar D$, Lemma 1.4 yields for $z \in D$

$$\tilde\gamma(z)$$
$$= \bar\partial\left\{-\int_{(V\setminus D)_\zeta} \bar\partial\tilde\gamma(\zeta)\wedge\mu^1_{q-1}\wedge\frac{1}{p!\,(n-p)!}D_{p,n-p}(\partial z,\partial\zeta) - I^2_{p,q-1}(V,\tilde\gamma)\right\}$$
$$-\int_{(V\setminus D)_\zeta} \bar\partial\tilde\gamma(\zeta)\wedge\bar\partial_\zeta\mu^2_q\wedge\frac{1}{p!\,(n-p)!}D_{p,n-p}(\partial z,\partial\zeta).$$

Since $\bar\partial\tilde\gamma = 0$ on $\partial(V\setminus D)$, the second integral above vanishes by Stokes' theorem. The first integral is a linear combination of integrals considered in Lemma 4.2; hence the proof is by appealing to Lemma 4.2 and Corollary 2.4.

COROLLARY 4.5. *If D is a strictly pseudoconvex domain in $\mathbf{C}^n$, $n \geqslant 2$, with boundary of class C^{m+1}, $m \geqslant 2$, and $g \in C^{m,\lambda}(C\bar D)$, and if $\bar\partial g$ can be extended to CD as a form of class $C^{m-1,\lambda}_{(0,1)}(CD)$, then g can be extended to $\mathbf{C}^n$ as a function in $C^m(\mathbf{C}^n)$.*

Also, for every extension of $\bar\partial_g$ to a form $\alpha \in Z^{m,\lambda}_{(0,1)}(\mathbf{C}^n)$, there exists an extension $f \in C^m(\mathbf{C}^n)$ of g such that $\bar\partial f = \alpha$.

PROOF. Suppose $\bar\partial g$ is extended to $\alpha_1 \in C^{m-1,\lambda}_{(0,1)}(\mathbf{C}^n)$. Then $\bar\partial\alpha_1 \in C^{m-2,\lambda}_{(0,2)}(\mathbf{C}^n)$, and $\operatorname{supp}\bar\partial\alpha_1 \subset \bar D$. Further, if $n = 2$, for any $\mu \in A_2(\bar D)$ we have

$$\int_D \mu\wedge\bar\partial\alpha_1 = \int_{D_1}\mu\wedge\bar\partial\alpha_1 = \int_{\partial D_1}\mu\wedge\alpha_1 = \int_{\partial D_1}\mu\wedge\bar\partial g = 0.$$

Here D_1 is a domain with smooth boundary such that $D \Subset D_1$ and $\mu \in A_2(D_1)$.

Now Theorem 4.3, III shows that there exists $h \in C^m(\mathbf{C}^n)$ such that $\bar\partial h = \alpha_1$ on CD. The function $\varphi = h - g$ is holomorphic on $C\bar D$; hence by Hartogs' theorem it can be extended to an entire function. Therefore $g = h - \varphi$ can be extended to $\mathbf{C}^n$ as a function of class $C^m(\mathbf{C}^n)$.

To prove the second part of Corollary 4.5, we note that, by Theorem 4.4, $\alpha = \bar\partial f_1$ with $f_1 \in C^m(\mathbf{C}^n)$. Now $g - f_1$ is holomorphic in CD; let f_2 be the entire function extending it to $\mathbf{C}^n$. Then $f = f_1 + f_2$ is the desired extension.

This corollary is a generalization of Hartogs' theorem to smooth functions. For domains which are not domains of holomorphy, an assertion similar to Corollary 2.5 is false. An example of a function g not extendible continuously to CD, for which $\bar\partial g$ admits a C^∞ extension to CD, will be constructed in §8 (see (8.1)).

CHAPTER II

FORMS ORTHOGONAL TO HOLOMORPHIC FORMS

§5. Polynomials orthogonal to holomorphic functions

1°. Let D be a bounded domain in $\mathbf{C}^n$, with smooth boundary, and let $0 \in D$. Let $w_k(z)$, $k = 1,\dots,n$, be C^1 functions on ∂D such that, for all $z \in \partial D$,

$$\langle w(z), z \rangle = 1. \tag{5.1}$$

We set([1])

$$\omega(w, z) = \sum_{k=1}^{n} (-1)^{k-1} w_k dw[k] \wedge dz,$$

where $w = (w_1,\dots,w_n)$ and $[k]$ signifies that dw_k is to be omitted. If $\omega(w, z)$ is nondegenerate on ∂D, then every $\alpha \in C_{2n-1}(\partial D)$ can be written as $\alpha = \varphi(z)\omega$, with $\varphi \in C(\partial D)$. Instead of discussing forms orthogonal (with respect to integration over ∂D) to holomorphic functions, we shall discuss functions φ that are orthogonal to holomorphic functions in the sense that

$$\int_{\partial D} f(z)\varphi(z)\omega(w, z) = 0 \tag{5.2}$$

for all $f \in A(\overline{D})$. We shall denote the subspace of such functions $\varphi \in C(\partial D)$ by $O(\partial D)$.

In certain situations, it is useful to have a description of the polynomials $P(z, w)$ which lie in $O(\partial D)$. Observe that, for $n = 1$, $P(z, w)$ and $\omega(w, z)$ are respectively $P(z, 1/z)$ and dz/z.

([1]) Note that $\omega(w, z) = \omega(w, z, 0)$ of §1.2°.

THEOREM 5.1. *$P(z,w)\in O(\partial D)$ if and only if*

$$\int_{\substack{\tau_1\geqslant 0,\dots,\tau_{n-1}\geqslant 0\\ \tau_1+\cdots+\tau_{n-1}\leqslant 1}} P\left(z_1,\dots,z_n,\frac{\tau_1}{z_1},\dots,\frac{\tau_{n-1}}{z_{n-1}},\frac{1-\tau_1-\cdots-\tau_{n-1}}{z_n}\right)d\tau_1\wedge\cdots\wedge d\tau_{n-1}$$

$$=\sum_{j=1}^{n} Q_j\left(z_j,z_1,\frac{1}{z_1},\dots,[j],\dots,z_n,\frac{1}{z_n}\right), \tag{5.3}$$

where the Q_j are polynomials such that $Q_j(0,z_1,\dots,1/z_n)=0$, $j=1,\dots,n$.

We divide the proof of this theorem into several lemmas.

LEMMA 5.2. *If $f\in A(\overline{D})$ and $P(z,w)$ has the form*

$$P(z,w)=\sum_{k=1}^{m} a_{\alpha^k,\beta^k} z^{\alpha^k} w^{\beta^k}, \tag{5.4}$$

where $\alpha^k=(\alpha_1^k,\dots,\alpha_n^k)$ and $\beta^k=(\beta_1^k,\dots,\beta_n^k)$, then

$$\int_{\partial D} P(z,w)f(z)\omega=(2\pi i)^n\sum_{k=1}^{m} a_{\alpha^k,\beta^k}\left[D^{\beta^k,0}\frac{f(z)z^{\alpha^k}}{(\|\beta^k\|+n-1)!}\right]_{z=0}. \tag{5.5}$$

This lemma follows easily from the Cauchy-Fantappiè formula (see (13.2)), written (for ζ sufficiently close to 0) in the form

$$f(\zeta)=\frac{(n-1)!}{(2\pi i)^n}\int_{\partial D}\frac{f(z)\omega(w,z)}{(1-\langle\zeta,w\rangle)^n}.$$

LEMMA 5.3. *In order that a polynomial $P(z,w)$ of the form* (5.4) *lie in $O(\partial D)$, it is necessary and sufficient that, for all n-tuples $\gamma=(\gamma_1,\dots,\gamma_n)$ of nonnegative integers,*

$$\sum_{k=1}^{m} a_{\alpha^k,\beta^k}\left[D^{\beta^k,0}\frac{z^{\alpha^k+\gamma}}{(\|\beta^k\|+n-1)!}\right]_{z=0}=0. \tag{5.6}$$

PROOF. It follows from (5.5) that (5.6) signifies orthogonality of $P(z,w)$ to all monomials z^γ; hence the necessity is obvious. The right side of (5.5) depends neither on the form of D, nor on the concrete choice of the $w_i(z)$, $i=1,\dots,n$. Thus, whether $P(z,w)$ lies in $O(\partial D)$ or not depends only on the properties of $P(z,w)$ itself. In particular, $P(z,w)\in O(\partial D)$ if and only if $P(z,w)\in O(\partial B(0,r))$, where $B(0,r)=\{z: |z_1|^2+\cdots+|z_n|^2<r^2\}$. We have seen in Examples 1 and 2 of 3° how the $w_i(z)$ can be conveniently chosen for a ball. Each $f\in A(\overline{B(0,r)})$ can be uniformly approximated on $\partial B(0,r)$ by polynomials in z. Hence the orthogonality of $P(z,w)$ to all monomials z^γ already implies that $P(z,w)\in O(\partial B(0,r))$.

Finally, from (5.6) we deduce

LEMMA 5.4. *A polynomial $P(z, w)$ of the form* (5.4) *lies in* $O(\partial D)$ *if and only if, for all n-tuples γ of nonnegative integers,*

$$\sum_{\beta^k-\alpha^k=\gamma} a_{\alpha^k,\beta^k}\frac{\beta_1^k!\cdots\beta_n^k!}{(\|\beta^k\|+n-1)!}=0. \tag{5.7}$$

The proof of Theorem 5.1 now reduces to comparing (5.7) with the equality

$$\int_{\substack{\tau_1\geqslant 0,\dots,\tau_{n-1}\geqslant 0\\ \tau_1+\cdots+\tau_{n-1}\leqslant 1}} P\left(z,\frac{\tau_1}{z_1},\dots,\frac{\tau_{n-1}}{z_{n-1}},\frac{1-\tau_1-\cdots-\tau_{n-1}}{z_n}\right)d\tau_1\wedge\cdots\wedge d\tau_{n-1}$$

$$=\sum_{\gamma}\left(\sum_{\beta^k-\alpha^k=\gamma} a_{\alpha^k,\beta^k}\frac{\beta_1^k!\cdots\beta_n^k!}{(\|\beta^k\|+n-1)!}\right)z^{-\gamma}.$$

COROLLARY 5.5. *Every polynomial $P(z, w)$ can be represented as the sum of a polynomial $Q(z, w)\in O(\partial D)$ and a polynomial $R(w)$ depending only on w.*

PROOF. It suffices to consider a polynomial $P(z, w)$ of the form (5.4) for which $\beta^k-\alpha^k=\gamma$ is a constant vector. If at least one coordinate of γ is negative, then $P\in O(\partial D)$. If all the coordinates of γ are nonnegative and (5.7) is satisfied, then again $P\in O(\partial D)$. It remains to consider the case when all the components of γ are nonnegative and (5.7) does not hold. In this case, we can add a monomial $a_{0,\gamma}w^\gamma$ to P, choosing $a_{0,\gamma}$ so that (5.7) is satisfied for the polynomial $P+a_{0,\gamma}w^\gamma$, i.e. that this "augmented" polynomial lies in $O(\partial D)$.

Note the inclusions $A_0(\partial D)\subset O(\partial D)\subset C(\partial D)$, where $A_0(\partial D)$ is the space of traces on ∂D of functions $f(z)$ holomorphic in D and continuous on $\overline{D}$ with $f(0)=0$. From Theorem 5.1 it follows that $O(\partial D)$ is not a ring for $n>1$. We stress that the characterization of the polynomials in $O(\partial D)$ given by Theorem 5.1 and the representation in Corollary 5.5 do not depend on the concrete form of D.

2°. For large classes of domains, the functions $w_1(z),\dots,w_n(z)$ can be chosen so that

$$\textit{polynomials } P(z,w) \textit{ are dense in } C(\partial D). \tag{5.8}$$

Then the forms $P(z, w)\omega$ are dense in $C_{2n-1}(\partial D)$. The following theorem may be considered as a new step in the characterization of forms orthogonal to holomorphic functions.

THEOREM 5.6. *If $w_i\in C^2(\partial D)$, $i=1,\dots,n$, then a polynomial $P(z, w)$ belongs to $O(\partial D)$ if and only if the form $P\omega$ can be extended into D as a $\bar\partial$-exact form of type $(n, n-1)$.*

For the proof, we need

LEMMA 5.7. *$P(z, w) \in O(\partial D)$ if and only if $P(z, w)$ can be represented as a linear combination of polynomials of the following two kinds.*

$$z^\alpha w^\beta, \quad \text{with } \alpha_k > \beta_k \text{ for some } k, \tag{5.9}$$

$$z^\alpha w^{\alpha+\gamma}(C_{\alpha+\gamma,l} - z_l w_l); \tag{5.10}$$

here α, β and γ are nonnegative integral vectors, and $C_{\beta,l} = (\beta_l + 1)(\|\beta\| + n)^{-1}$.

PROOF. It follows from Lemma 5.4 that polynomials of the form (5.9) or (5.10) lie in $O(\partial D)$. To prove the necessity, it is enough in view of the same lemma, to show that a polynomial

$$w^\gamma \sum_{k=1}^{m} a_{\alpha^k} z^{\alpha^k} w^{\alpha^k}, \tag{5.11}$$

lying in $O(\partial D)$ is a linear combination of polynomials of the form (5.10).

Without loss of generality, we may suppose that $\alpha^1 = 0$ and $\alpha^k_{l_k} \neq 0$, $k = 2, \ldots, m$. Then (5.11) can be written as

$$\sum_{k=2}^{m} a_{\alpha^k} \frac{z^{\alpha^k} w^{\alpha^k+\gamma}}{z_{l_k} w_{l_k}} \left(z_{l_k} w_{l_k} - C_{\alpha^k+\gamma,l_k}\right)$$
$$+ \sum_{k=2}^{m} a_{\alpha^k} C_{\alpha^k+\gamma,l_k} \frac{z^{\alpha^k} w^{\alpha^k+\gamma}}{z_{l_k} w_{l_k}} + a_0 w^\gamma. \tag{5.12}$$

The first term in (5.12) is a linear combination of the polynomials (5.10); the remaining polynomial has lower degree in z than the initial one and also lies in $O(\partial D)$. Repeating this argument, we will have (after finitely many steps) a representation of (5.11) as a sum of a linear combination of the polynomials (5.10) and a polynomial bw^γ; by Lemma 5.4, $bw^\gamma \in O(\partial D)$ only if $b = 0$.

PROOF OF THEOREM 5.6. Sufficiency follows immediately from Stokes' theorem. We shall prove the necessity for polynomials (5.9) and (5.10) (see Lemma 5.7).

Suppose the polynomial has the form (5.9). We may suppose that $k = 1$. Then

$$z^\alpha w^\beta = (z_1 w_1)^{\beta_1} z_2^{\alpha_2} \cdots z_n^{\alpha_n} w_2^{\beta_2} \cdots w_n^{\beta_n} z_1^{\alpha_1 - \beta_1}$$
$$= z_1^{\alpha_1-\beta_1}(1 - z_2 w_2 - \cdots - z_n w_n)^{\beta_1} z_2^{\alpha_2} \cdots z_n^{\alpha_n} w_2^{\beta_2} \cdots w_n^{\beta_n},$$

and so it is enough to consider the polynomial $P = z_1^{\alpha_1} \cdots z_n^{\alpha_n} w_2^{\beta_2} \cdots w_n^{\beta_n}$ with $\alpha_1 > 0$.

By Lemma 1.6, for $z_k \neq 0$ we have

$$\omega(w, z) = \frac{(n-1)!}{(2\pi i)^n}(-1)^{k-1}\frac{1}{z_k}dw[k] \wedge dz. \tag{5.13}$$

Therefore, for $z_1 \neq 0$

$$\begin{aligned} P\omega &= \frac{(n-1)!}{(2\pi i)^n} z_1^{\alpha_1 - 1} z_2^{\alpha_2} \cdots z_n^{\alpha_n} w_2^{\beta_2} \cdots w_n^{\beta_n} dw[1] \wedge dz \\ &= \bar\partial\left\{\frac{(n-1)!}{(2\pi i)^n(\beta_2+1)} z_1^{\alpha_1-1} z_2^{\alpha_2} \cdots z_n^{\alpha_n} w_2^{\beta_2+1} \cdots w_n^{\beta_n} dw[1,2] \wedge dz\right\}. \end{aligned}$$

This equality can be extended to all of ∂D by continuity.

It remains to consider the case when $P \in \mathrm{O}(\partial D)$ has the form (5.10). In this case

$$P(z, w) = z^\alpha w^{\alpha+\gamma}\left[C_{\alpha+\gamma, l}\sum_{k=1}^{n} z_k w_k - z_l w_l\right] = z^\alpha w^{\alpha+\gamma}\sum_{k=1}^{n} a_k z_k w_k.$$

For $z_1 \cdot \cdots \cdot z_n \neq 0$ we get from (5.13)

$$P\omega = \frac{(n-1)!}{(2\pi i)^n} z^\alpha w^{\alpha+\gamma}\sum_{m=1}^{n}(-1)^{m-1} a_m w_m dw[m] \wedge dz. \tag{5.14}$$

The condition $z_1 \cdot \cdots \cdot z_n \neq 0$ may be dropped since both sides of (5.14) are continuous on ∂D. If $m > 1$, then

$$\begin{aligned} &\bar\partial(w_1 w_m z^\alpha w^{\alpha+\gamma} dw[1, m] \wedge dz) \\ &\quad = (\alpha_1 + \gamma_1 + 1) z^\alpha w^{\alpha+\gamma} w_m dw[m] \wedge dz \\ &\quad\quad + (-1)^m(\alpha_m + \gamma_m + 1) z^\alpha w^{\alpha+\gamma} w_1 dw[1] \wedge dz, \end{aligned}$$

and so (5.14) implies

$$P\omega = a w_1 z^\alpha w^{\alpha+\gamma} dw[1] \wedge dz + \bar\partial_\mu,$$

where a is a constant, and μ a linear combination of the forms $w_1 w_m z^\alpha w^{\alpha+\gamma} dw[1, m] \wedge dz$. Further, in view of (5.13),

$$P\omega = \frac{(2\pi i)^n}{(n-1)!} a z_1 w_1 z^\alpha w^{\alpha+\beta}\omega + \bar\partial_\mu = P_1\omega + \bar\partial_\mu.$$

Hence $P_1(z, w) \in \mathrm{O}(\partial D)$, and it follows from (5.7) that $a = 0$.

3°. We now consider examples of classes of domains for which there exists a vector-valued function $w(z)$, $z \in \partial D$, such that (5.1) and (5.8) hold and the form $\omega(w, z)$ is nondegenerate.

EXAMPLE 1. Let $D = \{z\colon \rho(z) < 0\}$ be a strictly pseudoconvex bounded domain with C^2 boundary and containing zero; suppose that D is also linearly convex (Aĭzenberg [1]), which in this case means that, for every $z \in \partial D$, the analytic tangent plane $\{\zeta\colon (\zeta_1 - z_1)\rho'_{z_1} + \cdots + (\zeta_n - z_n)\rho'_{z_n} = 0\}$ does not intersect D. We set

$$w_i = \frac{\rho'_{z_i}}{z_1\rho'_{z_1} + \cdots + z_n\rho'_{z_n}}, \qquad i = 1,\ldots,n. \tag{5.15}$$

For this choice of w_i, (5.1) holds. Let us consider the form ω formed from the w_i as in 1°.

LEMMA 5.8. *$\omega(w, z)$ is nondegenerate on ∂D.*

PROOF. We represent ∂D as the union of disjoint sets $\Gamma_1,\ldots,\Gamma_n$ such that $\rho'_{\bar z_m} \neq 0$ on Γ_m. Then we see by an easy computation that, on Γ_m,

$$\omega = \frac{(-1)^{m-1}\mathfrak{L}(\rho)\,d\bar z[m]}{\rho'_{\bar z_m}(z_1\rho'_{z_1} + \cdots + z_n\rho'_{z_n})^n} \wedge dz,$$

where $\mathfrak{L}(\rho)$ is the generalization of the Levi determinant (see Fuks [1], Chapter II, §12.3, or Vladimirov [1], Chapter III, §18.5) to the case of n variables (see Rizza [1]):

$$\mathfrak{L}(\rho) = -\begin{vmatrix} 0 & \rho'_{z_1} & \cdots & \rho'_{z_n} \\ \rho'_{\bar z_1} & \rho''_{z_1,\bar z_1} & \cdots & \rho''_{z_n\bar z_1} \\ \cdots & \cdots & \cdots & \cdots \\ \rho'_{\bar z_n} & \rho''_{z_1\bar z_n} & \cdots & \rho''_{z_n\bar z_n} \end{vmatrix}.$$

It remains to show that $\mathfrak{L}(\rho) \neq 0$ on ∂D for strictly pseudoconvex domains D. But this can be easily checked by making a nonsingular linear change of variables such that the analytic tangent plane to D at a given point of ∂D is parallel to a coordinate hyperplane, and using the following property of $\mathfrak{L}(\rho)$: if $z = z(\zeta)$ is a biholomorphic map, and $\rho_1(\zeta) = \rho(z(\zeta))$, then

$$\mathfrak{L}(\rho_1) = \mathfrak{L}(\rho)\left|\frac{\partial(z_1,\ldots,z_n)}{\partial(\zeta_1,\ldots,\zeta_n)}\right|^2. \tag{5.16}$$

The proof of (5.16) in the general case is the same as in the case $n = 2$ (see Fuks [1], Chapter II, §12.3).

Let h be the locus of (z, w) in $\mathbf{C}^{2n}$ as z ranges over ∂D. As in the proof of Lemma 5.8 (by means of the same linear change of variables), it is easy to prove

LEMMA 5.9. *h is a smooth manifold without complex tangent vectors.*

Let $\tilde{D} = \{w: \langle w, z\rangle \neq 1 \text{ for all } z \in D\}$ be the compact set dual to D, and let $\tilde{\tilde{D}}$ be the dual of $\tilde{D}$. $\tilde{\tilde{D}}$ is an open set (see Aĭzenberg [1] and [2]). Let us also require that $\tilde{\tilde{D}}$ be connected.(²) Then D is linearly convex in the sense of Martineau [1].

LEMMA 5.10.

$$h = (\bar{D} \times \tilde{D}) \cap \{(z, w): \langle z, w\rangle = 1\}, \tag{5.17}$$

where h is a polynomially convex compact set.

PROOF. $\tilde{D}$ may be interpreted as the set of complex direction vectors w of analytic hyperplanes of the form $\alpha_w = \{z: \langle w, z\rangle = 1\}$ that do not meet D. Hence the right side of (5.17) is the set of pairs (z, w) in $\mathbf{C}^{2n}$ such that $z \in \partial D$ and α_w passes through z but does not meet D. From the uniqueness of the analytic tangent plane, it follows that w has the form (5.15). Thus (5.17) holds. $\bar{D}$ and $\tilde{D}$ are polynomially convex, since $\tilde{\tilde{D}}$ and $\tilde{\tilde{D}}$ are connected (see, for example, Makarova, Kudaĭberganov and Cherkashin [1]); hence (5.17) implies the polynomial convexity of h.

In view of Lemma 5.9, continuous functions on h are uniformly approximable by holomorphic functions (Harvey and Wells [1]). These in turn can be approximated by polynomials $P(z, w)$, because of Lemma 5.10. Thus condition (5.8) is satisfied.

EXAMPLE 2. Now, in contrast to (5.15), we set

$$w_j = \bar{z}_j/|z|^2, \qquad j = 1, \ldots, n. \tag{5.18}$$

For this choice of the w_j, the vector-valued function $w(z)$ does not depend on D, and always satisfies (5.1). Let us consider conditions under which (5.8) holds. Let D be a bounded domain with $0 \in D$. Let B denote the uniformly closed algebra of functions on ∂D generated by $z_1, \ldots, z_n$ and $w_1, \ldots, w_n$, and let B_α be its restriction to the (complex) one-dimensional analytic plane α, $0 \in \alpha$. Clearly, B separates points on ∂D and contains 1. Let $\alpha = \{z: z = at = (a_1 t, \ldots, a_n t), t \in \mathbf{C}^1\}$, $a \in \mathbf{C}^n$. Then $w_j = a_j |a|^{-2} t^{-1}$, so that B_α is generated by t and t^{-1}.

LEMMA 5.11. *For* (5.8) *to hold, it is necessary and sufficient that, for every* α, *the set* $\partial D \cap \alpha$ *have no interior points and that its complement in* α *consist of two connected components.*

For the proof, we need the following results of Čirka [1].

Let X be a compact set in $\mathbf{C}^n$, *a subalgebra of* $C(X)$ *and* $F = \{f_\alpha\}_{\alpha \in I}$ *a family of real-valued functions in* $C(X)$. *Then a continuous function f on X can be*

(²) It is not clear whether the connectedness of $\tilde{\tilde{D}}$ is a consequence of the hypotheses of Example 1.

approximated uniformly on X by polynomials in elements of A and F if and only if f can be approximated on every $X_\alpha = \{x \in X\colon f_\alpha(X) = a_\alpha\}$, $a_\alpha \in \mathbf{R}$, *by elements of A.*

PROOF OF LEMMA 5.11. Consider the functions $f_{jk} = \bar{z}_j z_k |z|^{-2}, j, k = 1, \ldots, n$, $f_{jk} = \bar{f}_{kj}$, $f_{jk} \in B$. Then $f_{jk} + f_{kj} = 2\operatorname{Re} f_{jk} \in B$. These functions do not separate points lying in the same plane. By taking the restriction of f_{jk} to two planes α_1 and α_2 and considering the functions $f_{jk} f_{jj}^{-1}$, it is easy to see that $\operatorname{Re} f_{jk}$ and $\operatorname{Im} f_{jk}$ separate points of different planes. Thus the set $\{z \in \partial D\colon \operatorname{Re} f_{jk} = b_{jk}, \operatorname{Im} f_{jk} = c_{jk}\}$ is either empty or coincides with $\partial D \cap \alpha$ for some α.

Sufficiency. If $\partial D \cap \alpha$ satisfies the conditions of the lemma, then, by a theorem of Mergeljan [1], polynomials in t and t^{-1} are dense in $C(\partial D \cap \alpha)$. Hence Čirka's result implies that $B = C(\partial D)$.

Necessity. If $B = C(\partial D)$, then

$$B_\alpha = C(\partial D \cap \alpha). \tag{5.19}$$

In the interior of $\partial D \cap \alpha$, each function in B_α, being a uniform limit of holomorphic functions, is holomorphic; hence (5.19) can hold only if $\partial D \cap \alpha$ has no interior points. If the complement of $\partial D \cap \alpha$ had a component ω containing neither 0 nor ∞, then for $t_0 \in \omega$ we would have $(t - t_0)^{-1} \in C(\partial D \cap \alpha)$; but $(t - t_0)^{-1} \notin B_\alpha$ by the maximum modulus principle.

It remains to consider the question of the nondegeneracy of the form ω for the choice (5.18) of $w(z)$.

LEMMA 5.12. *Let D be a bounded domain with smooth boundary. Then* $\omega(w, z)$ *is nondegenerate on* ∂D *if and only if, for all* $z \in \partial D$

$$z_1\rho'_{z_1} + \cdots + z_n\rho'_{z_n} \neq 0, \tag{5.20}$$

i.e. no analytic tangent plane passes through $0 \in D$.

The proof reduces to simple computation. For instance, if at $z \in \partial D$ we have $\rho'_{\bar{z}_n} \neq 0$, then

$$\begin{aligned}\omega(w, z) &= \frac{1}{|z|^{2n}} \sum_{k=1}^{n} (-1)^{k-1} \bar{z}_k \, d\bar{z}[k] \wedge dz \\ &= \frac{(-1)^n}{|z|^{2n}\rho'_{\bar{z}_n}} (z_1\rho'_{z_1} + \cdots + z_n\rho'_{z_n}) d\bar{z}[n] \wedge dz.\end{aligned}$$

Thus the nondegeneracy of ω is equivalent to (5.20).

Thus all the required conditions of 1° and 2° are satisfied by the function $w(z)$ of (5.18), provided D satisfies the conditions of Lemmas 5.11 and 5.12.

EXAMPLE 3. If the class of domains D is the same as in Examples 1 or 2, but $\rho \in C^{n+2}$, then, by Theorem 5.4 of Hörmander and Wermer [1], (5.8) is valid for polynomials in z and v, where $v = (v_1,\dots,v_n)$, each v_i being a C^{n+1} function sufficiently close to w_i in the C^2 topology, $i = 1,\dots,n$. Thus it remains only to choose the v_i such that the analogue of (5.1) holds and $\omega(v, z)$ is nondegenerate on ∂D.

EXAMPLE 4. Let D be a domain of the class of Example 1, and let $w_i \in C^{n+1}$ as in Example 3. We extend the w_i to some open neighborhood V of ∂D so that (5.1) is preserved. Consider homeomorphisms $\varphi_t \subset C^{n+1}$ of ∂D onto the boundary $\partial D_t \subset V$ of a domain D_t depending continuously in the C^2 topology on the parameter t, $0 < t < 1$, such that $\lim_{t\to 0}\varphi_t$ in the C^2 topology is the identity map of ∂D. Again applying Theorem 5.4 of Hörmander and Wermer, we see that, for $\varepsilon > 0$ sufficiently small, the polynomials $P(z, w)$ have the property (5.8) on ∂D_t, $0 < t < \varepsilon$. Thus the w_i of (5.15) work not only for the domains of Example 1, but also for domains which are close to them in the above sense.

§6. Forms orthogonal to holomorphic forms: the case of strictly pseudoconvex domains

1°. We now proceed to the characterization of forms orthogonal to holomorphic functions, i.e. we shall answer the following question:

Let D be a bounded domain with smooth boundary in $\mathbf{C}^n$. Then which forms $\alpha \in C_{(p,n-1)}(\partial D)$ are orthogonal to functions in $A_{n-p}(\overline{D})$ in the sense that

$$\int_{\partial D} \mu \wedge \alpha = 0$$

for all $\mu \in A_{n-p}(\overline{D})$?

Let us denote the space of these orthogonal forms by $A^{\perp}_{n-p}(\overline{D})$.

As already observed, in the case $n = 1$, for any domain D, we have $\alpha \in A^{\perp}_{1-p}(\overline{D})$ if and only if $\alpha \in Z^1_{(p,0)}(D) \cap C_{p,0}(\overline{D})$, $p = 0, 1$.

It is easy to give a sufficient condition for a form α to belong to $A^{\perp}_{n-p}(\overline{D})$. Indeed, if $\alpha = \gamma|_{\partial D}$ for some $\gamma \in Z^1_{(p,n-1)}(\overline{D})$, then $\alpha \in A^{\perp}_{n-p}(\overline{D})$, since

$$\int_{\partial D} \mu \wedge \alpha = \int_D d(\mu \wedge \gamma) = 0$$

by Stokes' theorem, for any $\mu \in A_{n-p}(\overline{D})$.

By a standard application of the Hahn-Banach theorem, it can be seen from Theorem 2.1 that $Z^{\infty}_{(n,n-1)}(\overline{D})$ is weakly dense in the space of forms (and even measures) from $AC^{\perp}(\overline{D})$ in the topology of $C^*(\partial D)$, the dual space of $C(\partial D)$.

2°. For strictly pseudoconvex domains, we can show that the condition stated in 1° is not only sufficient, but also necessary. More precisely, we have

THEOREM 6.1. *If D is a strictly pseudoconvex domain in $\mathbf{C}^n$, $n > 1$, with $\partial D \in C^{m+1,\lambda}$, $m \geqslant 1$, and if $\alpha \in C^{m,\lambda}_{(p,n-1)}(\partial D)$, then $\alpha \in A^{\perp}_{n-p}(\overline{D})$ if and only if there exists $\gamma \in C^m_{p,n-2}(\overline{D})$ such that $\bar\partial\gamma \,|_{\partial D} = \alpha$.*

Theorem 6.1, Part II of Theorem 4.3 and Part III of Theorem 4.3 for $q = n$ are essentially equivalent.

PROOF. Sufficiency was proved in 1°.

Necessity. By Theorem 2.12,

$$\alpha = \alpha_1 \,|_{\partial D} - \alpha_2 \,|_{\partial D}, \tag{6.1}$$

where $\alpha_1 \in Z^{m,\lambda'}_{(p,n-1)}(\overline{D})$ and $\alpha_2 \in Z^{m,\lambda'}_{(p,n-1)}(CD)$. If we extend α_2 to a form $\tilde\alpha_2 \in C^{m,\lambda'}_{(p,n-1)}(\mathbf{C}^n)$, then $\gamma = \bar\partial\tilde\alpha_2$ satisfies the conditions of Theorem 4.3:

$$\int_D \mu \wedge \gamma = \int_D \mu \wedge \bar\partial\tilde\alpha_2 = \int_{\partial D} \mu \wedge \alpha_2 = 0,$$

since $\alpha_2 \,|_{\partial D} = \alpha - \alpha_1 \,|_{\partial D}$, and $\alpha, \alpha_1 \,|_{\partial D} \in A^{\perp}_{n-p}(\overline{D})$. By Part III of Theorem 4.3, $\alpha_2 = \tilde\alpha_2 = \bar\partial\beta$ on CD, and $\beta \in C^m_{(p,n-2)}(\mathbf{C}^n)$.

By Theorem 4.4, $\alpha_1 = \bar\partial\gamma_1$ and $\gamma_1 \in C^m_{(p,n-2)}(\overline{D})$. Now (6.1) yields that $\gamma = \gamma_1 - \beta$ has the desired properties.

Theorem 6.1 cannot be carried over to arbitrary domains because it is not true that all $\bar\partial$-closed forms (which, as shown in 1°, are orthogonal to holomorphic forms) are $\bar\partial$-exact for every domain. Thus the following result seems natural:

THEOREM 6.2. *Let $D = \Omega \setminus \bigcup_1^k (\overline{\Omega}_i)$, where Ω and the Ω_i are strictly pseudoconvex domains with C^{m+2} boundary, $m \geqslant 1$, such that $\Omega_i \Subset \Omega$ and $\overline{\Omega}_i \cap \overline{\Omega}_j = \varnothing$ for $i = j$; let $\alpha \in C^{m,\lambda}_{(p,n-1)}(\partial D)$. Then $\alpha \in A^{\perp}_{n-p}(\overline{D})$ if and only if there exists $\tilde\alpha \in Z^{m-1}_{(p,n-1)}(\overline{D})$ such that $\tilde\alpha \,|_{\partial D} = \alpha$.*

The "exotic" form of the domain in which such a characterization is given reflects to a certain extent (and for $n = 2$ almost completely) the true nature of things, as Theorem 8.1 shows.

PROOF. Let $\alpha \in A^{\perp}_{n-p}(\overline{D})$. By Theorem 2.12, $\alpha = \alpha_1 \,|_{\partial D} - \alpha_2 \,|_{\partial D}$, where $\alpha_1 \in Z^{m,\lambda'}_{(p,n-1)}(\overline{D})$ and $\alpha_2 \in Z^{m,\lambda'}_{(p,n-1)}(CD)$. α_2 breaks up into forms $\alpha^0, \alpha^1, \ldots, \alpha^k$, $\alpha^0 \in Z^{m,\lambda'}_{(p,n-1)}(C\Omega)$, $\alpha_i \in Z^{m,\lambda'}_{(p,n-1)}(\Omega_i)$. By Theorem 4.4, $\alpha_i = \bar\partial\gamma_i$, $\gamma_i \in C^m_{(p,n-2)}(\Omega_i)$. Extend the γ_i to $\mathbf{C}^n$ so that

$$\operatorname{supp} \gamma_i \Subset \Omega \Big\backslash \bigcup_{\substack{j=1 \\ j \neq i}}^{k} \Omega_j,$$

and denote the extended forms also by γ_i. Consider $\beta = \alpha_2 - (\bar\partial\gamma_1 + \cdots + \bar\partial\gamma_k)$. Clearly, $\beta = 0$ on Ω_i, $i = 1, \ldots, k$, and $\beta = \alpha^0$ on $C\Omega$. Further,

$\beta|_{\partial D} \in A^{\perp}_{n-p}(\bar{D})$, since $\bar{\partial}\gamma_i|_{\partial D}$ and $\alpha_2|_{\partial D}$ belong to $A^{\perp}_{n-p}(\bar{D})$. And since $A_{n-p}(\bar{\Omega}) = A_{n-p}(\bar{D})$, we have $\alpha^0|_{\partial\Omega} \in A^{\perp}_{n-p}(\bar{\Omega})$. By Theorem 6.1, $\alpha^0|_{\partial D} = \bar{\partial}\gamma_0|_{\partial D}$, where $\gamma_0 \in C^m_{(p,n-1)}(\bar{\Omega})$. Let $\varphi \in C^\infty(\mathbf{C}^n)$, $\varphi \equiv 1$ on $C\Omega$ and $\varphi \equiv 0$ on $\bigcup_1^k \Omega_i$. Then $\beta|_{\partial D} = \bar{\partial}(\varphi\gamma_0)|_{\partial D}$. Set

$$\tilde{\alpha} = \alpha_1 - \sum_{i=1}^{k} \bar{\partial}\gamma_i - \bar{\partial}(\varphi\gamma_0).$$

Then $\tilde{\alpha} \in Z^{m-1}_{(p,n-1)}(\bar{D})$, and

$$\tilde{\alpha}|_{\partial D} = \alpha_1|_{\partial D} - \left(\sum_{i=1}^{k} \bar{\partial}\gamma_i|_{\partial D} + \bar{\partial}(\varphi\gamma_0)|_{\partial D}\right) = \alpha.$$

For $k = 1$, Theorem 6.2 admits an alternative formulation:

COROLLARY 6.3. *Let $D_1 \Subset D_2$ be strictly pseudoconvex domains with C^{m+2} boundaries, $m \geqslant 1$, and let $\alpha_i \in C^{m,\lambda}_{(p,n-1)}(\partial D_i)$, $i = 1, 2, \ldots$. Then the α_i are restrictions to the ∂D_i of the same form $\tilde{\alpha} \in Z^{m-1}_{(p,n-1)}(\bar{D}_2 \setminus D_1)$ if and only if*

$$\int_{\partial D_1} \mu \wedge \alpha_1 = \int_{\partial D_2} \mu \wedge \alpha_2$$

for any $\mu \in A_{n-p}(\bar{D}_2)$.

3°. Just as Parts II and III of Theorem 4.3 for $q = n$ yield Theorem 6.1, Parts I and III for $q < n$ yield

THEOREM 6.4. *If D is a strictly pseudoconvex domain in $\mathbf{C}^n$, $n \geqslant 2$, with C^{m+2} boundary, $m \geqslant 1$, and if $\alpha \in Z^{m,\lambda}_{(p,q)}(CD)$, $1 \leqslant q \leqslant n-2$, then there exists $\gamma \in C^m_{(p,q-1)}(\mathbf{C}^n)$ such that $\bar{\partial}\gamma = \alpha$ on CD.*

This theorem is an analogue of Hartogs' theorem on the extension of holomorphic functions from the exterior of a compact set.

§7. The general case

For the characterization of forms orthogonal to holomorphic forms in the general case, we need the notion of the envelope of holomorphy of a closed bounded domain. If a compact set K can be written as $\bigcap_m D_m$, $D_{m+1} \Subset D_m$, where each domain D_m has a schlicht (i.e. univalent) envelope of holomorphy $H(D_m)$, then we shall say that K has a *schlicht envelope of holomorphy*, and define $H(K)$ as $\bigcap_m H(D_m)$. The envelope of holomorphy $H(K)$ so obtained does not depend on the choice of the sequence D_m, $m = 1, 2, \ldots$, and preserves a number of properties of the envelopes of holomorphy of domains (see Aĭzenberg [2]).

THEOREM 7.1. *If* $\partial D \in C^{m+1,\lambda}$, $m \geqslant 1$, *a compact set* $\bar{D}$ *has a schlicht envelope of holomorphy, and* $\gamma \in C^{m,\lambda}_{(p,n-1)}(\partial D)$, *then* $\gamma \in A^{\perp}_{n-p}(\bar{D})$ *if and only if there exist* $\gamma_1 \in Z^{m,\lambda'}_{(p,n-1)}(\bar{D})$ *and* $\gamma_2 \in B_{p,n-1}(CD) \cap C^{m,\lambda'}_{(p,n-1)}(CD)$, $0 < \lambda' < \lambda$, *where* $\gamma_2 = O(|z|^{1-2n})$ *as* $|z| \to \infty$ *and is* $\bar{\partial}$*-exact in* $CH(D)$, *such that*

$$\gamma = \gamma_1 |_{\partial D} - \gamma_2 |_{\partial D}. \tag{7.1}$$

If $p = n$, *then* $\gamma_1 \in B_{n,n-1}(D)$.

PROOF. SUFFICIENCY. Consider $\alpha \in A_{n-p}(\bar{D})$. There exists a bounded domain Q with smooth boundary such that $H(\bar{D}) \subset Q$ and $\alpha \in A_{n-p}(\bar{Q})$. By assumption, $\gamma_2 = \bar{\partial}\beta$ in $CH(\bar{D})$. Now, by Stokes' theorem,

$$\int_{\partial D} \alpha \wedge \gamma = \int_{\partial D} \alpha \wedge \gamma_1 - \int_{\partial D} \alpha \wedge \gamma_2 = (-1)^{n-p} \int_D \alpha \wedge \bar{\partial}\gamma_1 - \int_{\partial Q} \alpha \wedge \bar{\partial}\beta$$

$$= (-1)^{n-p-1} \int_{\partial Q} d(\alpha \wedge \beta) = 0.$$

Necessity. Using Theorem 2.12, we write γ in the form (7.1). Obviously, $\gamma_1 \in A^{\perp}_{n-p}(\bar{D})$; hence $\gamma_2 \in A^{\perp}_{n-p}(\bar{D})$. We must show that γ_2 is $\bar{\partial}$-exact in $CH(\bar{D})$. Let $H(\bar{D}) = \bigcap Q_m$, where $Q_{m+1} \Subset Q_m$ and the Q_m are strictly pseudoconvex with C^∞ boundary. If $\alpha \in A_{n-p}(\bar{Q}_m)$, then, since γ_2 is $\bar{\partial}$-closed, we have by Stokes' theorem

$$\int_{\partial Q_m} \alpha \wedge \gamma_2 = \int_{\partial D} \alpha \wedge \gamma_2 = 0,$$

i.e. $\gamma_2 \in A^{\perp}_{n-p}(Q_m)$ for all m. As in the proof of Theorem 6.1, we get that for every m there exists $\delta_m \in C^\infty_{p,n-2}(\mathbf{C}^n)$ such that $\gamma_2 = \bar{\partial}\delta_m$ in CQ_m. By induction we shall now construct $\delta'_m \in C^\infty_{(p,n-2)}(CQ_m)$, $m = 1, 2, \ldots$, such that $\bar{\partial}\delta'_m = \gamma_2$ and $\delta'_m = \delta'_{m+1}$ on CQ_m.

We put $\delta'_1 = \delta_1$. Suppose δ'_m is already defined, and $\theta_m = \delta_{m+1} - \delta'_m$. Then $\bar{\partial}\theta_m = \bar{\partial}\delta_{m+1} - \bar{\partial}\delta'_m = 0$ on CQ_m. By Theorem 6.4, θ_m can be extended to all of $\mathbf{C}^n$ as a form in $Z^\infty_{(p,n-2)}(\mathbf{C}^n)$; denote this extension also by θ_m and set $\delta'_{m+1} = \delta_{m+1} - \theta_m$. Then $\bar{\partial}\delta'_{m+1} = \bar{\partial}\delta_{m+1} - \bar{\partial}\theta_m = \gamma_2$ on CQ_{m+1}, and $\delta'_{m+1} - \delta'_m = \delta_{m+1} - \theta_m - \delta'_m = 0$ on CQ_m.

If we define δ as equal to δ'_m on CQ_m, then $\delta \in C^\infty_{(p,n-2)}(CH(\bar{D}))$ and $\bar{\partial}\delta = \gamma_2$.

The question arises whether the second term on the right of (7.1) is necessary, or whether in the general case, as in the strictly pseudoconvex case, the characterizations of forms in $A^{\perp}_{n-p}(\bar{D})$ may be confined to that of forms with $\bar{\partial}$-closed extensions into D. From the results of §8, it follows that the second term in (7.1) is in general necessary.

§8. Converse theorems

It is natural to ask whether there exist domains for which it is impossible to characterize forms in $A_{n-p}^{\perp}(\bar{D})$ in a manner similar to Theorem 6.2, and if such domains do exist, then to describe the class of domains for which the characterization is possible.

For domains in $\mathbf{C}^2$, the answer is provided by the following result.

THEOREM 8.1. *Let D be a domain in $\mathbf{C}^2$ with C^2 boundary, and suppose that, for every $\alpha \in A^{\perp}(\bar{D}) \cap C^{\infty}_{(2,1)}(\partial D)$, there exists $\tilde{\alpha} \in Z^1_{(2,1)}(\bar{D})$ such that $\tilde{\alpha}|_{\partial D} = \alpha$. Then*

$$D = \omega \setminus \bigcup_{i=1}^{k} \bar{\Omega}_i,$$

where Ω and the Ω_i are domains of holomorphy, $\Omega_i \Subset \Omega$, and $\bar{\Omega}_i \cap \bar{\Omega}_j = \varnothing$ for $i \neq j$.

PROOF. Since D is a bounded domain with boundary of class C^2, we have $D = \Omega \setminus \bigcup_1^k \Omega_i$, where Ω and the Ω_i are domains with C^2 boundary and connected complements, with $\Omega_i \Subset \Omega$ and $\bar{\Omega}_i \cap \bar{\Omega}_j = \varnothing$.

1. Suppose Ω is not a domain of holomorphy. Then there exists $z^0 \in \partial D$ at which the Levi form is negative definite; hence there is a neighborhood U of z_0 such that $\{z \in U\colon F(z, z_0) = 0\} \setminus z^0 \subset \Omega$, where

$$F(z, z^0) = \sum_{i=1}^{2} \frac{\partial \rho}{\partial z_i}(z^0)(z_i - z_i^0) + \sum_{i,j=1}^{2} \frac{\partial^2 \rho}{\partial z_i \partial z_j}(z^0)(z_i - z_i^0)(z_j - z_j^0)$$

(see, for example, Gunning and Rossi [1], Chapter IX, §B).

Let $\varphi \in C^{\infty}(\mathbf{C}^2)$, and $\operatorname{supp} \varphi \Subset U$ and $\varphi \equiv 1$ in a neighborhood of z^0. Put

$$g(z) = \varphi(z)\left[F(z, z^0)\right]^{-1} \tag{8.1}$$

for $z \in C\bar{\Omega}$. Then $\bar{\partial} g \in Z^{\infty}_{(0,1)}(C\Omega)$, since $\bar{\partial} g = 0$ at $z \notin U \cup \Omega$ and $\bar{\partial} g = \bar{\partial}\varphi(z)[F(z, z^0)]^{-1}$ for $z \in U \setminus \Omega$. Further, if

$$\alpha = \begin{cases} \bar{\partial} g \wedge dz\,|_{\partial\Omega} & \text{on } \partial\Omega, \\ 0 & \text{on } \partial\Omega_i,\ i = 1, \dots, k, \end{cases}$$

then $\alpha \in A^{\perp}(\bar{D})$. Indeed, α satisfies the conditions of Theorem 7.1, since $\alpha = \alpha_1|_{\partial D} - \alpha_2|_{\partial D}$, where $\alpha_1 \equiv 0$, and $\alpha_2 = \bar{\partial} g \wedge dz$ on $C\Omega$ and 0 on the Ω_i. By assumption there is an $\tilde{\alpha} \in Z^1_{(2,1)}(\bar{D})$ such that $\tilde{\alpha}|_{\partial D} = \alpha$.

By (1.1) applied to $B\backslash\Omega$, where B is any ball such that $U \cup \Omega \Subset B$, we have for $z \in B\backslash\overline{\Omega}$

$$\bar\partial g \wedge dz = \int_{[\partial(B\backslash\Omega)]_\zeta} \bar\partial g(\zeta) \wedge d\zeta \wedge U_{2,1}(\zeta, z) - \bar\partial \int_{(B\backslash\Omega)_\zeta} \bar\partial g(\zeta) \wedge d\zeta \wedge U_{2,0}(\zeta, z). \tag{8.2}$$

By the same formula applied to D, we have for $z \in B\backslash\overline{\Omega}$

$$0 = \int_{(\partial D)_\zeta} \tilde\alpha(\zeta) \wedge U_{2,1}(\zeta, z) - \bar\partial \int_{D_\zeta} \tilde\alpha(\zeta) \wedge U_{2,0}(\zeta, z). \tag{8.3}$$

The first integral of (8.2) and that of (8.3) are negatives of each other, since $\bar\partial g \,|_{\partial B} = \tilde\alpha \,|_{\partial\Omega_i} = 0$ and $\tilde\alpha \,|_{\partial\Omega} = \bar\partial g \wedge dz \,|_{\partial\Omega}$. Hence by adding (8.2) and (8.3) we get $\bar\partial g \wedge dz = \bar\partial h \wedge dz$, where h is defined by

$$h(z) \wedge dz = -\int_{(B\backslash\Omega)_\zeta} \bar\partial g(\zeta) \wedge d\zeta \wedge U_{2,0}(\zeta, z) - \int_{D_\zeta} \tilde\alpha \wedge U_{2,0}(\zeta, z). \tag{8.4}$$

$h(z)$ is continuous in $\mathbf{C}^2$, since the integrands in (8.4) have integrable singularities (see (1.4)). Now $g - h$ is holomorphic in $B\backslash\Omega$, since $\bar\partial(g - h) \wedge dz = 0$, so that $g - h$ extends holomorphically into B. Therefore g extends continuously to z_0. On the other hand,

$$\lim_{z \to z^0} |g(z)| = \infty.$$

This contradiction shows that the assumption that Ω is not a domain of holomorphy is false.

2. Suppose now that some Ω_i is not a domain of holomorphy. Then there is a ball B such that all functions of $A(\Omega_i)$ extend holomorphically to B, and $B\backslash\overline{\Omega}_i$ is not empty. Let $\zeta^0 \in B\backslash\overline{\Omega}_i$. Without loss of generality, we may suppose that $\zeta^0 = 0$ and $B \cap \Omega_i \wedge \{z \in \mathbf{C}^2 : z_2 = 0\} \neq \varnothing$.

$$\alpha = \begin{cases} U_{2,1}(0, z) \,|_{\partial\Omega_i} & \text{on } \partial\Omega_i, \\ 0 & \text{on } \partial\Omega \text{ and } \Omega_j, j \neq i. \end{cases}$$

Exactly as in step 1 above, we see that $\alpha \in A^{\perp}(\bar{D})$. Hence there is an $\tilde{\alpha} \in Z^1_{(2,1)}(\bar{D})$ such that $\tilde{\alpha}|_{\partial D} = \alpha$. By (1.1) applied to Ω_i, for $z \in \Omega_i$ we get

$$U_{2,1}(0, z) = \int_{(\partial \dot{\Omega}_i)_\zeta} U_{2,1}(0, \zeta) \wedge U_{2,1}(\zeta, z)$$

$$-\bar{\partial} \int_{(\dot{\Omega}_i)_\zeta} U_{2,1}(0, \zeta) \wedge U_{2,0}(\zeta, z). \tag{8.5}$$

By the same formula applied to D, for $z \in \Omega_i$ we get

$$0 = \int_{\partial D_\zeta} \tilde{\alpha}(\zeta) \wedge U_{2,1}(\zeta, z) - \bar{\partial} \int_D \tilde{\alpha}(\zeta) \wedge U_{2,0}(\zeta, z). \tag{8.6}$$

Again as in step 1, we see that the first integrals of (8.5) and (8.6) are negatives of each other. Hence, if we add them together, (8.5) and (8.6) yield $U_{2,1}(0, z) = \bar{\partial} h \wedge dz$, with h defined by

$$h(z) \wedge dz = -\int_{(\dot{\Omega}_i)_\zeta} U_{2,1}(0, \zeta) \wedge U_{2,0}(\zeta, z) - \int_{D_\zeta} \tilde{\alpha}(\zeta) \wedge U_{2,0}(\zeta, z). \tag{8.7}$$

By Corollary 2.4, the first integral on the right above is of class $C^{\infty}_{(2,0)}(\Omega_i)$ and the second integral is proper for $z \in \Omega_i$, so that it also lies in $C^{\infty}_{(2,0)}(\Omega_i)$, i.e. $h \in C^{\infty}(\Omega_i)$.

On the other hand,

$$U_{2,1}(0, z) = \frac{1}{(2\pi i)^2} \frac{\bar{z}_1 d\bar{z}_2 - \bar{z}_2 d\bar{z}_1}{|z|^4} \wedge dz = \bar{\partial} \frac{\bar{z}_1 dz}{(2\pi i)^2 |z|^2 z_2}$$

for $z_2 \neq 0$; hence $h - \bar{z}_1((2\pi i)^2 |z|^2 z_2)^{-1}$ is holomorphic in $D \backslash \{z\colon z_2 = 0\}$. Then $g = z_2 h - \bar{z}_1((2\pi i)^2 |z|^2)^{-1}$ is holomorphic in D, and so it extends holomorphically to B. But $g(z_1, 0) = z_1^{-1}$ and $B \cap \{z_2 = 0\}$ is connected; hence g cannot be extended to $0 \in B$. The theorem is proved in full.

Corollary 8.2. *Let D be a domain in $\mathbf{C}^2$ with C^2 boundary, and suppose that, for each $\alpha \in A^{\perp}(\bar{D}) \cap C^{\infty}_{(2,1)}(\partial D)$, there exists $\gamma \in C^1_{(2,0)}(\bar{D})$ such that $\bar{\partial}\gamma \in Z^1_{(2,1)}(\bar{D})$ and $\partial\gamma|_{\partial D} = \alpha$. Then D is a domain of holomorphy.*

Proof. By Theorem 8.1,

$$D = \Omega \Big\backslash \bigcup_{i=1}^{k} \bar{\Omega}_i,$$

where Ω is a domain of holomorphy and k is the number of bounded components of CD. The corollary will be proved if we show that $k = 0$. Suppose that $k \neq 0$, and let ζ^0 lie in a bounded component of CD, and let Γ be the boundary of this component. Then $U_{2,1}(\zeta^0, \zeta) \in Z^{\infty}_{(2,1)}(\bar{D})$ by Lemma 1.2;

hence $U_{2,1}(\zeta^0, \zeta)\,|_{\partial D} \in A^{\perp}(\overline{D})$. By hypothesis, we therefore have a $\gamma \in C^1_{(2,0)}(\overline{D})$ such that $U_{2,1}(\zeta^0, z) = \bar{\partial}\gamma\,|_{\partial D}$. But then by the Martinelli–Bochner formula ((1.1) for $p = q = 0$) and Stokes' theorem we get

$$1 = \int_\Gamma U_{2,1}(\zeta^0, \zeta) = \int_\Gamma \bar{\partial}\gamma = 0.$$

This contradiction proves the corollary.

CHAPTER III

PROPERTIES OF $\bar{\partial}$-CLOSED FORMS OF TYPE $(p, n-1)$

§9. The theorems of Runge and Morera

The classical theorem of Runge asserts that a domain in $\mathbf{C}^1$ is polynomially convex if and only if it is simply connected. Wermer (see Wermer [1] or Gunning and Rossi [1]) has constructed an example of a domain which is biholomorphically equivalent to the tricylinder but not polynomially convex. Thus there are no necessary and sufficient topological conditions on a pair of domains in $\mathbf{C}^n$ which ensure that homolorphic functions in the smaller domain can be approximated by functions which are holomorphic in the larger domain. For $\bar{\partial}$-closed forms of type $(p, n-1)$, the situation is exactly the same as that for holomorphic functions of one complex variable.

THEOREM 9.1 (Runge's theorem for forms). *Let $\Omega_1 \subset \Omega_2$ be open sets in $\mathbf{C}^n$. Then the following conditions are equivalent*:

1. *Every form lying in $Z^m_{(p,n-1)}(\Omega_1)$ can be approximated in $C^m_{(p,n-1)}(\Omega_1)$ by forms lying in $Z^\infty_{(p,n-1)}(\Omega_2)$.*

2. *If $\Omega_2 \setminus \Omega_1 = K \cap F$ with K compact and F closed in Ω_2, and if $K \cap F = \varnothing$, then $K = \varnothing$.*(1)

PROOF. $1 \to 2$. Suppose $\Omega_2 \setminus \Omega_1 = K \cup F$, with F closed in Ω_2, K compact, $K \cap F = \varnothing$ but $K \neq \varnothing$. Pick $\zeta^0 \in K$. Then

$$\alpha = D_{1,n-1}\big(t(\zeta, \zeta^0), \bar{\partial}_\zeta t(\zeta, \zeta^0)\big) \wedge d\zeta_1 \wedge \cdots \wedge d\zeta_p \in Z^\infty_{(p,n-1)}(\Omega_1).$$

By condition 1 we can approximate α in $C^0_{(p,n-1)}(\Omega_1)$ by the $\alpha_k \in Z^\infty_{(p,n-1)}(\Omega_2)$. Let D be an open set, consisting of finitely many connected components with

(1) Condition 2 means precisely that each bounded component of $C\Omega_1$ meets $C\Omega_2$.

smooth boundary, such that $K \subset D \Subset \Omega_2 \setminus F$. Then $\partial D \subset \Omega_1$, and by Stokes' theorem

$$\int_{\partial D} \alpha \wedge d\zeta_{p+1} \wedge \cdots \wedge d\zeta_n = \lim_{k\to\infty} \int_{\partial D} \alpha_k \wedge d\zeta_{p+1} \wedge \cdots \wedge d\zeta_n$$
$$= \lim_{k\to\infty} \int_D \bar\partial \alpha_k \wedge d\zeta_{p+1} \wedge \cdots \wedge d\zeta_n = 0.$$

On the other hand, by the definition of $U_{0,0}$ and the Martinelli-Bochner formula (see (1.1)), we have

$$\int_{\partial D} \alpha \wedge d\zeta_{p+1} \wedge \cdots \wedge d\zeta_n = \frac{(2\pi i)^n}{(n-1)!} \int_{\partial D} U_{0,0}(\zeta, \zeta^0) = \frac{(2\pi i)^n}{(n-1)!}.$$

This contradiction proves that $1 \to 2$.

We shall prove the implication $2 \to 1$ first only for $m = \infty$; the case of an arbitrary m will be handled later, using Theorem 9.3.

By the Hahn-Banach theorem, it is sufficient to prove that any continuous linear function T on $C^\infty_{(p,n-1)}(\Omega_1)$ which vanishes on $Z^\infty_{(p,n-1)}(\Omega_2)$ vanishes on $Z^\infty_{(p,n-1)}(\Omega_1)$.

Now, since T is linear and continuous on $C^\infty_{(p,n-1)}(\Omega_1)$, which is the projective limit of the spaces $C^\infty_{(p,n-1)}(L)$, T can be extended to a continuous linear functional on $C^\infty_{(p,n-1)}(L)$ for some compact set $L \Subset \Omega_1$ (see, for example, Gel'fand and Šilov [1], Chapter I, §4.1). Here we may assume that CL has no bounded connected components contained in Ω_1; otherwise we need only adjoin to L all the bounded components of CL contained in Ω_1 to obtain a new compact set which does satisfy the required condition.

Put

$$\psi(\zeta) = T_z\big(U_{p,n-1}(\zeta, z)\big), \qquad \zeta \in CL. \tag{9.1}$$

For a C^∞ function g, $\Delta g/\Delta x_k$ tends to g'_{x_k} uniformly on all compact sets; hence we may differentiate (9.1) with respect to ζ and use Lemma 1.2 to get

$$\bar\partial\psi = T_z\big(\bar\partial_\zeta U_{p,n-1}(\zeta, z)\big) = (-1)^n T_z\big(\bar\partial_z U_{p,n-2}(\zeta, z)\big)$$
$$= (-1)^n T_z\big(\bar\partial_z(\varphi_\zeta(z) U_{p,n-2}(\zeta, z))\big) = 0,$$

where $\varphi_\zeta \in C^\infty(\mathbf{C}^n)$ is equal to 1 in a neighborhood of L, and equal to 0 in a neighborhood of $\zeta \in CL$. This means that $\psi(\zeta)$ is holomorphic on CL.

For $\zeta \in C\Omega_2$,

$$D^{I,0}\psi = T_z\big(D^{I,0}_\zeta U_{p,n-1}(\zeta, z)\big) = 0,$$

since $U_{p,n-1}(\zeta, z)$, and hence all its derivatives, belong to $Z^\infty_{(p,n-1)}(\Omega_2)$ (for $\zeta \in C\Omega_2$). By the uniqueness theorem, $\psi \equiv 0$ in each component of CL

meeting $C\Omega_2$. Also, $\psi \equiv 0$ in the unbounded component of CL as well. This is proved for $n = 1$ in Hörmander [1] (§1.3). For $n > 1$, ψ can be holomorphically extended from the unbounded component of CL to all of $\mathbf{C}^n$ by Hartogs' theorem, and

$$\lim_{|\zeta| \to \infty} \psi(\zeta) = 0.$$

From Liouville's theorem it follows that $\psi \equiv 0$ in the unbounded component of CL. Condition 2 of the theorem and the choice of L guarantee that CL has no other components, i.e. $\psi \equiv 0$ on CL.

Now let $\alpha \in Z^{\infty}_{(p,n-1)}(\Omega_1)$, and let D be an open set with smooth boundary consisting of finitely many connected components $L \subset D \Subset \Omega_1$. Let $\varphi \in C^{\infty}(\mathbf{C}^n)$, $\varphi \equiv 1$ in a neighborhood U of L, and $\operatorname{supp} \varphi \Subset D$. Then, for $z \in U$, (1.1) gives

$$\begin{aligned} \alpha(z) &= I^1_{p,n-1}(D, \alpha) - \bar\partial I^2_{p,n-2}(D, \alpha) \\ &= \int_{(\partial D)_\zeta} \alpha(\zeta) \wedge U_{p,n-1}(\zeta, z) - \bar\partial\left(\varphi I^2_{p,n-2}(D, \alpha)\right). \end{aligned}$$

Since $\varphi I^2_{p,n-2}(D, \alpha) \in C^{\infty}_{(p,n-2)}(\mathbf{C}^n)$ and $\partial D \subset CL$, we deduce that

$$\begin{aligned} T\alpha &= \int_{\partial D_\zeta} T_z\left(\alpha(\zeta) \wedge U_{p,n-1}(\zeta, z)\right) - T\left(\bar\partial\left(\varphi I^2_{p,n-2}(D, \alpha)\right)\right) \\ &= \int_{\partial D} \alpha(\zeta) \wedge \psi(\zeta) = 0. \end{aligned}$$

Thus $2 \to 1$ is proved (for $m = \infty$).

As in the case of one variable (see Hörmander [1], §1.3), we have, in particular, for $\Omega_2 = \mathbf{C}^n$

COROLLARY 9.2. *$Z^{\infty}_{(p,n-1)}(\mathbf{C}^n)$ is dense in $Z^m_{(p,n-1)}(\Omega)$ if and only if Ω has connected complement in the one-point compactification of $\mathbf{C}^n$.*

THEOREM 9.3. *Let Ω be an open set in $\mathbf{C}^n$ and $\alpha \in C^m_{(p,n-1)}(\Omega)$, $m \geqslant 0$. Then the following conditions are equivalent:*

1. *For every ball $B \Subset \Omega$ and every $I = (i_1, \ldots, i_{n-p})$, $1 \leqslant i_1 < \cdots < i_p \leqslant n$,*

$$\int_{\partial B} \alpha \wedge d\zeta_I = 0. \tag{9.2}$$

2. *α can be approximated in $C^m_{(p,n-1)}(\Omega)$ by forms from $Z^{\infty}_{(p,n-1)}(\Omega)$.*
3. *$\alpha \in Z^m_{(p,n-1)}(\Omega)$.*

Weierstrass' theorem on the limit of a sequence of holomorphic functions shows that, for $n = 1$, Theorem 9.3 is precisely Morera's theorem. For $n > 1$, a form satisfying the equivalent conditions of Theorem 9.3 can be nondifferentiable. As an example of such a form we may take $\varphi(z_1)d\bar{z}[n]$, where $\varphi(z_1)$ is a continuous but not differentiable function on $\mathbf{C}^1$. Thus it is impossible to obtain a theorem in the same formulation as Morera's theorem for $n = 1$.

PROOF OF THEOREM 9.3. We first introduce some notation.

If g is locally integrable in Ω and $f \in C^\infty(\mathbf{C}^n)$ with compact support L, then for all $z \in \Omega$ such that $z - L \subset \Omega$ (2) we set

$$(g * f)(z) = \int_{\mathbf{C}^n_\zeta} g(\zeta)f(z-\zeta)d\bar{\zeta} \wedge d\zeta$$

$$= \int_{\mathbf{C}^n_\zeta} g(z-\zeta)f(\zeta)d\bar{\zeta} \wedge d\zeta.$$

$$\rho_r(z) = \begin{cases} \dfrac{K}{r^n} \exp\left(-\dfrac{r^2}{r^2 - |z|^2}\right) & \text{for } |z| < r, \\ 0 & \text{for } |z| \geqslant r \end{cases} \tag{9.3}$$

(K is chosen so that $\int \rho_r \, d\bar{\zeta} \wedge d\zeta = 1$).

If

$$\alpha = \sum_J{}' \sum_{k=1}^n a_{kJ} d\bar{z}[k] \wedge dz_J \in C^m_{(p,n-1)}(\Omega),$$

then

$$\alpha_r(z) = \sum_J{}' \sum_{k=1}^n (a_{kJ} * \rho_r)(z) d\bar{z}[k] \wedge dz_J.$$

It is known (see, for example, Bremermann [1], §3.2, Lemma 2) that $\alpha_r \in C^\infty_{(p,n-1)}(\Omega^r)$, where $\Omega^r = \{z \in \Omega: \rho(z, \partial\Omega) > r\}$, and that α_r converges to α in $C^m_{(p,n-1)}(\Omega)$ as $r \to \infty$.

We first prove that $1 \to 2$. Let $B \Subset \Omega$, and let the multi-index I' be such that $dz_{I'} \wedge dz_I = dz$. Then

$$\int_B \bar\partial\alpha_r \wedge dz_I = \int_{\partial B} \alpha_r \wedge dz_I$$

$$= \int_{\partial B_z} \sum_{k=1}^n \int_{\mathbf{C}^n_\zeta} a_{kI'}(z-\zeta)\rho_r(\zeta) d\bar{\zeta} \wedge d\zeta \wedge d\bar{z}[k] \wedge dz$$

$$= \int_{\mathbf{C}^n_\zeta} \left(\int_{\partial B_z} \alpha(z-\zeta) \wedge dz_I \right) \rho_r(\zeta) d\bar{\zeta} \wedge d\zeta$$

$$= \int_{\mathbf{C}^n_\zeta} \left(\int_{(\partial B - \zeta)_z} \alpha(z) \wedge dz_I \right) \rho_r(\zeta) d\bar{\zeta} \wedge d\zeta = 0.$$

(2) $z - L$ denotes the vector difference, i.e. the set $\{\zeta \in \mathbf{C}^n: \zeta = z - w, w \in L\}$.

Hence $\bar\partial\alpha_r \wedge dz_I \equiv 0$ in Ω^r. Since the multi-index I is arbitrary, it follows that $\bar\partial\alpha_r = 0$. Now it is easy to see that the open sets Ω^r, Ω satisfy condition 2 of Theorem 9.1. Thus we may apply Theorem 9.1 with $m = \infty$ to conclude that $1 \to 2$.

For $m > 0$, the implication $3 \to 1$ is an immediate consequence of Stokes' theorem. So, let $m = 0$ and $B \Subset \Omega$. Then $B \Subset \Omega^r$ for sufficiently small r, since B is compact, and so

$$\int_{\partial B} \alpha \wedge dz_I = \lim_{r\to 0} \int_{\partial B} \alpha_r \wedge dz_I = \lim_{r\to 0} \int_B \bar\partial\alpha_r \wedge dz_I$$

$$= \lim_{r\to 0} \int_{B_z} \sum_{k=1}^{n} (-1)^{k-1} \left(\int_{\mathbf{C}^n_\zeta} a_{kI'}(\zeta) \frac{\partial}{\partial \bar z_k} \rho_r(z-\zeta)\, d\bar\zeta \wedge d\zeta \right) d\bar z \wedge dz,$$

where I' is as before. Using the equality $[\rho_r(z-\zeta)]'_{\bar z_k} = -[\rho_r(z-\zeta)]'_{\bar\zeta_k}$ and the definition of a derivative in the space of distributions, we get

$$\int_{\partial B} \alpha \wedge dz_I = \lim_{r\to 0} \int_{B_z} d\bar z \wedge dz \int_{\mathbf{C}^n_\zeta} \sum_{k=1}^{n} (-1)^{k-1} \frac{\partial}{\partial \bar\zeta_k} a_{kI'}(\zeta) \rho_r(z-\zeta)\, d\bar\zeta \wedge d\zeta$$

$$= \lim_{r\to 0} \int_{\mathbf{C}^n_\zeta} d\bar\zeta \wedge d\zeta \int_{B_z} \bar\partial\alpha(z)\rho_r(z-\zeta) \wedge dz_I = 0.$$

This proves $3 \to 1$.

The implication $2 \to 3$ is trivial for $m \geqslant 1$, and follows for $m = 0$ from the fact that differentiation is a continuous operator in the space of distributions.

CONCLUSIONS OF THE PROOF OF THEOREM 9.1. By Theorem 9.3, every $\alpha \in Z^m_{(p,n-1)}(\Omega_1)$ can be approximated in $C^m_{(p,n-1)}(\Omega_1)$ by forms from $Z^\infty_{(p,n-1)}(\Omega_1)$. Hence Theorem 9.1 for the case $m = \infty$ can be applied to deduce the theorem in the general case.

§10. The first Cousin problem, separation of singularities and domains of existence

For forms of type $(p, n-1)$, we can pose the following problem, analogous to the Mittag-Leffler problem:

Let Ω be an open set in $\mathbf{C}^n$ and $\{z^i\}$ a discrete sequence of distinct points in Ω. Suppose forms $\alpha_i \in Z^\infty_{(p,n-1)}(\Omega\setminus\{z^i\})$ are given. Does there exist an $\alpha \in Z^\infty_{(p,n-1)}(\Omega\setminus\{z^i\})$ such that $\alpha - \alpha_i \in Z^\infty_{(p,n-1)}(U_i)$ for some neighborhood U_i of z^i?

As in the case of functions, this problem can be generalized to the following one, which it is natural to call the first Cousin problem for forms:

Let Ω be an open set in $\mathbf{C}^n$, and $\{\Omega_i\}$ an open covering of Ω. Suppose forms $\alpha_{ij} \in Z^\infty_{(p,n-1)}(\Omega_i \cap \Omega_j)$ are given, satisfying the conditions

$$\alpha_{ij} + \alpha_{ji} = 0 \quad \text{on } \Omega_i \cap \Omega_j,$$

$$\alpha_{ij} + \alpha_{jk} + \alpha_{ki} = 0 \quad \text{on } \Omega_i \cap \Omega_j \cap \Omega_k$$

for all i, j and k. Can we find forms $\alpha_i \in Z^\infty_{(p,n-1)}(\Omega_i)$ such that $\alpha_{ij} = \alpha_j - \alpha_i$ on $\Omega_i \cap \Omega_j$?

The forms α_{ij} are called the *Cousin data*, and the α_i (if they exist) a *solution of the first Cousin problem.*

LEMMA 10.1. *For any $\beta \in C^m_{(p,n)}(\Omega)$, the equation $\bar\partial\gamma = \beta$ has a solution $\gamma \in C^m_{(p,n-1)}(\Omega)$.*

PROOF. Let $\{r_k\}$ be a sequence of positive numbers decreasing to 0, such that $G_k = \Omega^{r_k} \cap B(0, k) \neq \varnothing$, $k = 1, 2, \ldots$, and let $\psi_k \in C^\infty(\mathbf{C}^n)$ be equal to 1 on G_{k+1} and $\operatorname{supp} \psi_k \Subset \Omega$. Put $\varphi_1 = \psi_1$, and $\varphi_k = \psi_k - \psi_{k-1}$ for $k > 1$. Then $\varphi_k = 0$ on G_k and $\varphi_1 + \cdots + \varphi_l + \cdots = 1$ on Ω. By (1.1) we have $\bar\partial\gamma_k = \varphi_k\beta$, where

$$\gamma_k = \int_{\Omega_\zeta} \varphi_k \beta \wedge U_{p,n-1}(\zeta, z).$$

By Corollary 2.4, $\gamma_k \in C^m_{(p,n-1)}(\Omega)$. Since $\varphi_k \equiv 0$ on G_k, $k > 1$, we get $\gamma_k \in Z^m_{(p,n-1)}(G_k)$. As we observed in the proof of Theorem 9.3, Ω^{r_k} and Ω satisfy condition 2 of Theorem 9.1; hence so do G_k and Ω. Thus we can choose $\chi_k \in Z^\infty_{(p,n-1)}(\Omega)$, $k > 1$, such that the differences between the corresponding coefficients of γ_k and χ_k, and all their derivatives of order up to $\min(m, k)$, are less (in absolute value) than 2^{-k} on G_{k-1} since $G_{k-1} \Subset G_k$. Then the series

$$\gamma = \gamma_1 + \sum_{k=2}^{\infty} (\gamma_k - \chi_k)$$

together with all its derivatives of order up to m converges uniformly in the interior of Ω to the form γ, so that $\gamma \in C^m_{(p,n-1)}(\Omega)$. By termwise differentiation, we get

$$\bar\partial\gamma = \varphi_1\beta + \sum_{k=2}^{\infty} \varphi_k\beta = \beta.$$

THEOREM 10.2. *If the Cousin data are of class C^m, $m \geqslant 0$, then the first Cousin problem has a solution which is of class C^m.*

Note that for the solvability, with any data, of the Cousin problem for holomorphic functions of several complex variables, it is necessary and sufficient that the first Dolbeault cohomology group be trivial. Therefore Theorem 10.2 is not valid as it stands in the case of holomorphic functions of several variables.

PROOF OF THEOREM 10.2. Let $\{\varphi_l\}$ be a partition of unity subordinate to the covering $\{\omega_i\}$, i.e. the sequences $\{\varphi_l\}$ of functions and $\{i_l\}$ of integers satisfy the following conditions:

1. $\varphi_l \in C^\infty(\mathbf{C}^n)$ and $\operatorname{supp} \varphi_l \Subset \Omega_{i_l}$.
2. On any compact subset of Ω, all but finitely many of the φ_l vanish.
3. $\sum_1^\infty \varphi_l \equiv 1$ on Ω.

Put $h_k = \sum_1^\infty \varphi_l \alpha_{i_l k}$; then $h_k \in C^m_{(p,n-1)}(\Omega_k)$. Furthermore,

$$h_k - h_j = \sum_{l=1}^{\infty} \varphi_l(\alpha_{i_l k} - \alpha_{i_l j}) = \sum_{l=1}^{\infty} \varphi_l \alpha_{jk} = \alpha_{jk},$$

and so the form β which is equal to $\bar\partial h_k$ on Ω_k is well-defined. And $\beta \in C^m_{(p,n)}(\Omega)$, since

$$\beta\,|_{\Omega_k} = \bar\partial h_k = \sum_{l=1}^{\infty} \bar\partial\varphi_l \wedge \alpha_{i_l k}.$$

By Lemma 10.1, there exists $\gamma \in C^m_{(p,n-1)}(\Omega)$ such that $\bar\partial\gamma = \beta$. The forms $\alpha_k = h_k - \gamma$ are the desired ones, since $\bar\partial\alpha_k = \bar\partial h_k - \bar\partial\gamma = 0$ and $\alpha_k - \alpha_j = h_k - \gamma - h_j + \gamma = \alpha_{jk}$.

In the special case when the covering consists of two sets, we have

COROLLARY 10.3. *Let Ω, Ω_1 and Ω_2 be open sets in $\mathbf{C}^n$, with $\Omega = \Omega_1 \cap \Omega_2$. Then, for any $\alpha \in Z^m_{(p,n-1)}(\Omega)$, there exist $\alpha_i \in Z^m_{(p,n-1)}(\Omega_i)$, $i = 1, 2$, such that $\alpha = \alpha_1 - \alpha_2$ on Ω.*

That is, for $\bar\partial$-closed $(p, n-1)$-forms, the theorem on separation of singularities is valid for any triple of domains, whereas for holomorphic functions of several complex variables this is not always the case.

THEOREM 10.4. *For any domain $D \subset \mathbf{C}^n$ and any p, $0 \leqslant p \leqslant n$, there exists a form $\alpha \in B_{p,n-1}(D)$ which cannot be continued across ∂D as a $\bar\partial$-closed form.*

PROOF. Let $\{\zeta^k\}$ be a countable dense subset of ∂D. Consider the function

$$G(z) = \sum_{k=1}^{\infty} \frac{1}{k^2} \cdot \frac{1}{|\zeta^k - z|^{2n-2}}. \tag{10.1}$$

If $K \subset D$ is compact, then $|\zeta - z| \geqslant r > 0$ for $\zeta \in \partial D$, $z \in K$ and an r independent of ζ and z. Therefore, the series (10.1) is majorized by the convergent series

$$\frac{1}{r^{2n-2}} \sum_{k=1}^{\infty} \frac{1}{k^2}.$$

Hence (10.1) converges uniformly in the interior of D to a harmonic function, and

$$\lim_{z \to z^0 \in \partial D} G(z) = \infty. \tag{10.2}$$

Clearly, the form

$$\alpha = \sum_J{}' \sum_{k=1}^{n} \frac{\partial G}{\partial z_k} d\bar z[k] \wedge dz_J,$$

has the desired properties (here $J = (j_1, \ldots, j_p)$). Indeed, suppose that α can be extended continuously to a neighborhood of a point $z^0 \in \partial D$. Then all the functions G'_{z_k}, and hence also all the $G'_{\bar z_k} = \overline{(G'_{z_k})}$,(³) can be extended continuously to a neighborhood of z^0. But then G itself can be extended smoothly to this neighborhood, which is impossible in view of (10.2).

§11. Theorems of approximation on compact sets

$\bar\partial$-closed $(p, n-1)$-forms enjoy approximation properties analogous to those of holomorphic functions of one complex variable on compact sets just as well as on open sets:

THEOREM 11.1. *Let K be a compact set in $\mathbf{C}^n$, and $\alpha \in C^m_{(p,n-1)}(K)$, $m \geqslant 1$. Suppose that α has an extension to a neighborhood of K such that $\bar\partial\alpha$ and all derivatives of order up to $m-1$ vanish on K. Then α can be appoximated in $C^{m-1}_{(p,n-1)}(K)$ by forms which are C^∞ and $\bar\partial$-closed in a neighborhood of K.*

If in addition CK is connected, then the approximating forms may be chosen in $Z^\infty_{(p,n-1)}(C^n)$.

If formulated in the same way as above for holomorphic functions of several complex variables, Theorem 11.1 would no longer be true (Fuks [2], Chapter III, §16.3).

PROOF. Let D be a bounded open set, consisting of finitely many connected components with smooth boundary, such that $K \subset D$ and $\alpha \in C^m_{(p,n-1)}(\bar D)$.

We represent α as an integral by means of (1.1). By Lemma 1.2,

$$I^1_{p,n-1}(D, \alpha) \in Z^\infty_{(p,n-1)}(D).$$

(³) This equality holds because G is real-valued.

By Corollary 2.4,

$$I^2_{p,n-2}(D,\alpha) \in C^m_{(p,n-2)}(D),$$

so we can approximate $I^2_{p,n-2}(D,\alpha)$ in the topology of $C^m_{p,n-2}(D)$ by forms lying in $C^\infty_{(p,n-2)}$. Then $\bar\partial I^2_{p,n-2}(\alpha)$ is approximated in $C^{m-1}_{(p,n-1)}(D)$ by forms lying in $Z^\infty_{(p,n-1)}(D)$.

To prove the theorem, it remains to approximate $I^2_{p,n-1}(D,\bar\partial\alpha)$. Now $\bar\partial\alpha$ has the form

$$\sum_J{}' h_J d\bar z \wedge dz_J, \qquad J=(j_1,\ldots,j_p),\ h_J \in C^{m-1}(\bar D).$$

Here the h_J vanish with all derivatives up to order $m-1$ on K; hence they can be approximated in $C^{m-1}(\bar D)$ by functions $h^l_J \in C^\infty(\bar D)$ which vanish in a neighborhood of K (see, for example, Malgrange [1], Chapter I, Lemma 4.2). Then

$$\alpha^l = I^2_{p,n-1}\Big(D, \sum_J{}' h^l_J d\bar\zeta \wedge d\zeta\Big) \in C^\infty_{(p,n-1)}(D).$$

The α^l are $\bar\partial$-closed in a neighborhood of K (since the h^l_J vanish in a neighborhood of K), and in this neighborhood we may differentiate the integral defining the operator $I^2_{p,n-1}$ under the integral sign.

We shall prove by induction on m that the sequence $\{\alpha^l\}$ converges in $C^{m-1}_{(p,n-1)}(K)$ to $\alpha^0 = I^2_{p,n-1}(D,\bar\partial\alpha)$.

First let $m=1$. For each J, $\{h^l_J\}$ converges uniformly on D to h_J, i.e., given $\varepsilon>0$, we can find an N such that $|h^l_J - h_J| < \varepsilon$ on $\bar D$ for $l>N$. Let α^l_{kJ} denote the coefficient of $d\bar z[k] \wedge dz_J$ in α^l, $l=0,1,\ldots$. Then (see (1.14))

$$\alpha^l_{kJ}(z) = \frac{(n-1)!}{(2\pi i)^n}\int_{D_\zeta} h^l_J(\zeta)(-1)^{k-1}\frac{\bar\zeta_k - \bar z_k}{|\zeta - z|^{2n}} d\bar\zeta \wedge d\zeta, \tag{11.1}$$

for $l=0,1,\ldots$ ($h^0_J = h_J$). Hence for $z\in K$ and $l>N$

$$\begin{aligned}|\alpha^l_{kJ}(z) - \alpha^0_{kJ}(z)| &= \left|\frac{(n-1)!}{(2\pi i)^n}\int_{D_\zeta}(h^l_J(\zeta)-h_J(\zeta))(-1)^{k-1}\frac{\bar\zeta_k-\bar z_k}{|\zeta-z|^{2n}}d\bar\zeta\wedge d\zeta\right| \\ &\leqslant \frac{\varepsilon(n-1)!}{\pi^n}\int_D \frac{dv}{|\zeta - z|^{2n-1}} \leqslant C\varepsilon,\end{aligned}$$

where

$$C = \frac{(n-1)!}{\pi^n}\int_\Omega \frac{dv}{|\zeta|^{2n-1}}, \qquad \Omega = \bigcup_{z\in K}(z+D),$$

and dv denotes Lebesgue measure.

Suppose now that our assertion has been proved for $m = s$. We proceed to prove it for $m = s + 1$. Note that $h_J \equiv 0$ in a neighborhood of K, and that the kernel in (11.1) is a function of $\zeta - z$. Hence, by Stokes' theorem, for $z \in K$ we have

$$\frac{\partial}{\partial z_i}\alpha^l_{kJ}(z) = \int_{D_\zeta} \frac{\partial}{\partial \zeta_i} h^l_J(\zeta)(-1)^{k-1} \frac{\bar{\zeta}_k - \bar{z}_k}{|\zeta - z|^{2n}} d\bar{z} \wedge d\zeta$$
$$- \int_{\partial D_\zeta} h^l_J(\zeta)(-1)^{k+i} \frac{\bar{\zeta}_k - \bar{z}_k}{|\zeta - z|^{2n}} d\bar{\zeta} \wedge d\zeta[i]. \qquad (11.2)$$

A similar formula is also valid for the derivatives with respect to the $\bar{z}_i$. The sequence formed by the second integral in (11.2) converges, with all its derivatives, uniformly on K; hence we see by means of the induction hypothesis that the sequence $\{(\alpha^l_{kJ})'_{z_i}\}$ converges in $C^{s-1}(K)$. Therefore $\{\alpha^l_{kJ}\}$ converges in $C^s(K)$ to α^0_{kJ}, and the first assertion of the theorem is proved.

Now, if CK is connected, then we may also suppose that the neighborhoods of K in which the approximating forms are defined also have connected complements. Then the second assertion follows from Theorem 9.1.

We shall call a domain D in $\mathbf{C}^n$ *admissible* if, for every $z \in \partial D$, there exists a neighborhood U_z of z and an $a \in \mathbf{C}^n$ such that

$$\bar{D} \cap U_z + \varepsilon a \Subset D \qquad (11.3)$$

for $0 < \varepsilon < 1$.

All domains whose boundaries have continuously varying normals are admissible; in particular, all domains with smooth boundaries are admissible.

For an admissible domain D, we define a function $r(z)$ on ∂D by $r(z) = \sup\{r > 0\colon$ there exists $a \in \mathbf{C}^n$ such that (11.3) holds for a and $U_z = B(z, r)\}$.

By the definition of an admissible domain, $r(z) > 0$. It is not hard to show that $r(z)$ is lower semicontinuous.

THEOREM 11.2. *Let D be an admissible bounded domain, and let $\alpha \in Z_{(p,n-1)}(\bar{D})$. Then α can be approximated in $C_{(p,n-1)}(\bar{D})$ by infinitely differentiable $\bar{\partial}$-closed forms defined in a neighborhood of $\bar{D}$.*

Further, if $\alpha \in Z^k_{(p,n-1)}(D)$, then the approximating sequence converges to α in $C^k_{(p,n-1)}(D)$.

For the proof, we need

LEMMA 11.3. *Let $D = D_1 \cap D_2$ (D, D_1 and D_2 are domains in $\mathbf{C}^n$) and $(D_1 \setminus \bar{D}) \cap (D_2 \setminus \bar{D}) = \varnothing$. Then any $\alpha \in Z_{(p,n-1)}(\bar{D})$ can be represented in the form $\alpha = \alpha_1 - \alpha_2$, where $\alpha_i \in Z_{(p,n-1)}(\bar{D}_i)$, $i = 1, 2$.*

If, moreover, $\alpha \in C^k_{(p,n-1)}$, then $\alpha_i \in C^k_{(p,n-1)}(D_i)$, $i = 1, 2$.

PROOF. The hypothesis of the lemma implies that $\overline{D} = \overline{D}_1 \cap \overline{D}_2$. Choose φ_1 and φ_2 in $C^\infty(\mathbf{C}^n)$ so that $\varphi_1 - \varphi_2 \equiv 1$ and $\operatorname{supp} \varphi_i \cap \overline{D_i \setminus \overline{D}} = \varnothing$. ($\varphi_1$ and φ_2 exist. Indeed, choose neighborhoods U_1 and U_2 of $D_1 \setminus \overline{D}$ and $D_2 \setminus \overline{D}$ such that $\overline{U}_1 \cap \overline{U}_2 = \varnothing$, and functions ψ_i such that $\psi_i = 0$ on U_i and $\psi_i > 0$ on $C\overline{U}_i$, $\psi_i \in C^\infty(\mathbf{C}^n)$. The $\varphi_1 = \psi_1(\psi_1 + \psi_2)^{-1}$ and $\varphi_2 = -\psi_2(\psi_1 + \psi_2)^{-1}$ will serve.)

We define two forms

$$\beta(z) = \begin{cases} \bar\partial \varphi_1 \wedge \alpha & \text{on } \overline{D}_1, \\ \bar\partial \varphi_2 \wedge \alpha & \text{on } \overline{D}_2, \end{cases}$$

$$\gamma(z) = \int_{(D_1 \cup D_2)_\zeta} \beta(\zeta) \wedge U_{p,n-1}(\zeta, z).$$

β is well-defined since $\bar\partial \varphi_1 = \bar\partial \varphi_2$ on $\overline{D}_1 \cap \overline{D}_2 = \overline{D}$, and we have $\beta \in C_{(p,n)}(\overline{D}_1 \cup \overline{D}_2)$ and $\gamma \in C_{(p,n-1)}(\overline{D}_1 \cup \overline{D}_2)$. We shall prove that

$$\beta = \bar\partial \gamma. \tag{11.4}$$

Here, the differentiation is to be understood in the sense of distributions.

Approximate β in $C_{(p,n)}(\overline{D}_1 \cup \overline{D}_2)$ by forms $\beta_i \in C^\infty_{(p,n)}(\overline{D}_1 \cup \overline{D}_2)$, and set

$$\gamma_i(z) = \int_{(D_1 \cup D_2)_\zeta} \beta_i(\zeta) \wedge U_{p,n-1}(\zeta, z).$$

By (1.1) we have $\beta_i = \bar\partial \gamma_i$, and since the γ_i approximate γ in $C_{(p,n-1)}(\overline{D}_1 \cup \overline{D}_2)$, (11.4) follows from the continuity of the differentiation operator in the space of distributions.

We set $\alpha_i = \varphi_i \alpha - \gamma$, $i = 1, 2$. Then $\alpha_i \in Z_{(p,n-1)}(\overline{D}_i)$, since $\bar\partial \alpha_i = \bar\partial \varphi_i \wedge \alpha - \bar\partial \gamma = \beta - \beta = 0$. Further, $\alpha_1 - \alpha_2 = \varphi_1 \alpha - \gamma - \varphi_2 \alpha + \gamma = (\varphi_1 - \varphi_2)\alpha = \alpha$ on $\overline{D}$. If moreover $\alpha \in C^k_{(p,n-1)}(D)$, then $\beta \in C^k_{(p,n)}(D_1 \cup D_2)$ by Corollary 2.4. Then $\gamma \in C^k_{(p,n-1)}(D_1 \cup D_2)$, and so $\alpha_i \in C^k_{(p,n-1)}(D_i)$.

PROOF OF THEOREM 11.2. Let $\alpha \in Z_{(p,n-1)}(\overline{D})$. Then for every $\delta > 0$ there exist domains D_i, $D \subset D_i$, and forms $\alpha_i \in Z_{(p,n-1)}(\overline{D}_i)$, $i = 1, \ldots, l$, such that

$$\alpha_1 + \cdots + \alpha_l = \alpha \tag{11.5}$$

on D and $\operatorname{diam}(\partial D \cap \partial D_i) < \delta$. We prove this as follows. Choose N such that $(2 \cdot 3^{-1})^N \operatorname{diam} D < \delta (2n)^{1/2}$. In the first step, put $b = \inf_{z \in \partial D} \operatorname{Re} z_1$, $c = \sup_{z \in \partial D} \operatorname{Re} z_1$ and $R = \sup_{z \in D} |z| + 1$. Then the domains D,

$$D^1 = D \cup \{z\colon |z| < R, \operatorname{Re} z_1 < b + (c - b)/3\}$$

and

$$D^2 = D \cup \{z\colon |z| < R, \operatorname{Re} z_1 > c - (c - b)/3\}$$

satisfy the conditions of Lemma 11.3. Hence $\alpha = \alpha^1 + \alpha^2$, where $\alpha^i \in Z_{(p,n-1)}(\overline{D^i})$. Observe that

$$\sup_{z,\zeta\in\partial D\cap\partial D^i} |\operatorname{Re}(\zeta_1 - z_1)| = \tfrac{2}{3}(c-b) \leqslant \tfrac{2}{3}\operatorname{diam} D.$$

At the next step, we put

$$b_i = \inf_{z\in\partial D\cap\partial D^i} \operatorname{Re} z_1, \qquad c_i = \sup_{z\in\partial D\cap\partial D^i} \operatorname{Re} z_1, \qquad i = 1, 2,$$

and apply Lemma 11.3 to the triples consisting of the domains D^i,

$$D^{i1} = D^i \cup \{z \colon |z| < R,\ \operatorname{Re} z_1 < b_i + (c_i - b_i)/3\}$$

and

$$D^{i2} = D^i \cup \{z \colon |z| < R,\ \operatorname{Re} z_1 > c_i - (c_i - b_i)/3\}.$$

We obtain forms $\alpha^{ij} \in Z_{(p,n-1)}(\overline{D^{ij}})$ with $\alpha = \alpha^{11} + \alpha^{12} + \alpha^{21} + \alpha^{22}$. Furthermore,

$$\sup_{z,\zeta\in\partial D\cap\partial D^{ij}} |\operatorname{Re}(\zeta_1 - z_1)| = \tfrac{2}{3} \sup_{z,\zeta\in\partial D\cap\partial D^i} |\operatorname{Re}(\zeta_1 - z_1)| \leqslant \left(\tfrac{2}{3}\right)^2 \operatorname{diam} D.$$

After N steps, we get 2^N domains $D^{i_1\cdots i_N}$ containing D and 2^N forms $\alpha^{i_1\cdots i_N} \in Z_{(p,n-1)}(\overline{D^{i_1\cdots i_N}})$ such that

$$\alpha = \sum_{i_1\cdots i_N=1}^{2} \alpha^{i_1\cdots i_N}$$

and

$$\sup_{z,\zeta\in\partial D\cap\partial D^{i_1\cdots i_N}} |\operatorname{Re}(\zeta_1 - z_1)| \leqslant \left(\frac{2}{3}\right)^N \operatorname{diam} D < \frac{\delta}{\sqrt{2n}}.$$

We repeat this process $2n-1$ more times, replacing $\operatorname{Re} z_1$ first by $\operatorname{Im} z_1$, then by $\operatorname{Re} z_2$, and so on. We obtain domains $D_1, \ldots, D_l$ and forms $\alpha_i \in Z_{(p,n-1)}(D_i)$ satisfying (11.5), with $\partial D \cap \partial D_i$ contained in a cube of edge $\delta(2n)^{-1/2}$. Hence $\operatorname{diam}(\partial D \cap \partial D_i) < \delta$.

We see from (11.5) that it suffices to prove the theorem for forms $\alpha \in Z_{(p,n-1)}(\overline{G})$, where G is a domain such that $D \subset G$ and the diameter δ of $\partial D \cap \partial G$ is small. Since $r(z)$ is lower semicontinuous and positive, there exists $\theta > 0$ such that $r(z) > \theta$. Let $\delta < \theta$, and fix a z^0 in $\partial D \cap \partial G$. Then there exist $r > \delta$ and $a \in \mathbf{C}^n$ such that $\overline{D} \cap B(z_0, r) + \varepsilon a \Subset D$ for $0 < \varepsilon < 1$. Since $\partial D \cap \partial G \subset B(z^0, r)$, we have $\overline{D} \setminus B(z^0, r) \Subset \overline{D}$; hence sufficiently small translations do not carry it outside G. Hence $\alpha_\varepsilon(z) = \alpha(z + \varepsilon a) \in Z_{(p,n-1)}(U_\varepsilon(\overline{D}))$

($U_\varepsilon(\bar{D})$ being a neighborhood of $\bar{D}$) for sufficiently small ε. As $\varepsilon \to 0$, the α_ε converge to α in $C_{(p,n-1)}(\bar{D})$. Applying Theorem 9.3 to the forms $\alpha_\varepsilon(z)$ and the domains $U_\varepsilon(\bar{D})$, we thus obtain the first assertion of Theorem 11.2.

Suppose now that in addition $\alpha \in C^k_{(p,n-1)}(D)$. Then, by the second part of Lemma 11.3, we see that $\alpha_i \in C^k_{(p,n-1)}(D_i)$. Thus we need only note that α_ε tends to α in $C^k_{(p,n-1)}(D)$ as $\varepsilon \to 0$.

THEOREM 11.4. *Suppose K is a compact set of $2n$-dimensional measure zero in $\mathbf{C}^n$. Then the space of forms of type $(p, n-1)$ which are C^∞ and $\bar\partial$-closed in a neighborhood of K is dense in $C_{(p,n-1)}(K)$.*

PROOF. By the Weierstrass approximation theorem, $C^\infty_{(p,n-1)}(\mathbf{C}^n)$ is dense in $C_{(p,n-1)}(K)$. Let $\alpha \in C^\infty_{(p,n-1)}(\mathbf{C}^n)$, and let B be a ball containing K. By (1.1),

$$\alpha = I^1_{p,n-1}(B, \alpha) - I^2_{p,n-1}(B, \bar\partial\alpha) - \bar\partial I^2_{p,n-2}(B, \alpha). \tag{11.6}$$

The first and third forms on the right above lie in $Z^\infty_{(p,n-1)}(B)$. Choosing a sequence of domains D_i such that $D_i \Subset D_{i-1} \subset B$ and $\cap D_i = K$, we have

$$I^2_{p,n-1}(B, \bar\partial\alpha)(z) = \lim_{i\to\infty} \int_{(B\setminus D_i)_\zeta} \bar\partial\alpha(\zeta) \wedge U_{p,n-1}(\zeta, z),$$

since the $2n$-dimensional measure of K is zero; the convergence is uniform in $z \in K$. Now the theorem follows from the fact that

$$\int_{(B\setminus D_i)_\zeta} \bar\partial\alpha(\zeta) \wedge U_{p,n-1}(\zeta, z) \in Z^\infty_{(p,n-1)}(D_i).$$

COROLLARY 11.5. *If D is a bounded domain in $\mathbf{C}^n$ with piecewise smooth boundary and z^0 an arbitrary point of D, then forms of the type $\alpha_1 + \alpha_2$, where $\alpha_1 \in Z^\infty_{(p,n-1)}(\bar{D})$ and $\alpha_2 \in Z^\infty_{(p,n-1)}(\mathbf{C}^n\setminus z^0)$, are dense in $C_{(p,n-1)}(\partial D)$.*

PROOF. ∂D satisfies the conditions of Theorem 11.4. Hence forms of type $(p, n-1)$ which are C^∞ and $\bar\partial$-closed in a neighborhood of ∂D are dense in $C_{(p,n-1)}(\partial D)$. Thus we need only approximate forms of that kind. Let ω be an open set containing ∂D, and let $\alpha \in Z^\infty_{(p,n-1)}(\omega)$. Choose a connected compact set $K \subset D$ so that $z^0 \in K$ and $\bar{D} \subset K \cup \omega$. By Corollary 10.4, $\alpha = \alpha_1 + \tilde\alpha_2$ on $\omega\setminus K$, where $\alpha_1 \in Z^\infty_{(p,n-1)}(\omega \cup D)$ and $\tilde\alpha_2 \in Z^\infty_{(p,n-1)}(CK)$. Since K is connected and compact, CK and $\mathbf{C}^n\setminus z^0$ satisfy condition 2 of Theorem 9.1. Hence $Z^\infty_{(p,n-1)}(\mathbf{C}^n\setminus z^0)$ is dense in $Z^\infty_{(p,n-1)}(CK)$, and the corollary follows.

CHAPTER IV

SOME APPLICATIONS

§12. Generalization of the theorems of Hartogs and F. and M. Riesz

1°. Let D be a bounded domain with piecewise smooth boundary, for which there exists a sequence of domains D_m with smooth boundaries satisfying the following conditions:

1) $D_m \Subset D_{m+1}$, and $\bigcup D_m = D$.

2) For every holomorphic function $f(z)$ in D the integral

$$\int_{\partial D_m} |f| dv_m \tag{12.1}$$

is nondecreasing as m increases, dv_m being the $(2n-1)$-dimensional Lebesgue volume element on ∂D_m.

3) If the integrals (12.1) are uniformly bounded for $f(z)$ (in which case we shall say $f(z) \in H_1$), then $f|_{\partial D_m}$ converges weakly to a summable function on ∂D (also denoted by $f(z)$, $z \in \partial D$) in the sense that

$$\lim_{m\to\infty} \int_{\partial D_m} |f| dv_m \leq \int_{\partial D} |f| dv; \qquad \lim_{m\to\infty} \int_{\partial D_m} f\nu = \int_{\partial D} f\nu$$

for every exterior differential form $\nu \in C_{2n-1}(\overline{D} \setminus D_1)$.

For $n = 1$, the theory of the class H_1 has been worked out in detail (Privalov [1], Hoffman [1]). For $n > 1$, the class H_1 can be constructed by using boundary properties of holomorphic functions of several complex variables. We indicate another possible approach. Suppose D is such that each of its sections by a complex one-dimensional analytic plane through 0 is simply connected, and that there is a C^1 homeomorphism ψ of D onto the unit balls under which each plane section of D through 0 is mapped conformally onto the section of the ball by the same plane.(¹) Let D_m be the inverse image by ψ of

(¹) These two conditions are satisfied, for instance, for linearly convex domains with smooth boundary (see Yuzhakov and Krivokoleskio [1]).

the ball of radius $1-(m+1)^{-1}$, $m=1,2,\ldots$. Then these D_m satisfy the conditions listed at the beginning of the section. If $f \in H_1$, then, for almost all plane sections, $f \in H_1$ in the one-variable sense (see Aĭzenberg [3], [4]), and the function $f(z)|_{\partial D}$ can be constructed from the sections utilizing the known boundary properties of holomorphic functions in H_1 for $n=1$.

Consider a smooth vector-valued function $w(z)$ on ∂D for which (5.1) holds and the form $\omega(w, z)$ is nondegenerate on ∂D. In addition to the space $\mathrm{O}(\partial D)$ introduced in §5, we shall need two of its subspaces. Let $\mathrm{O}_1(\partial D)$ be the subspace of $\varphi \in C(\partial D)$ such that there exists $\beta \in Z^{\infty}_{(n,n-1)}(\overline{D})$ for which $\beta|_{\partial D} = \varphi\omega$, and let $\mathrm{O}_2(\partial D)$ be the space of φ in $C(\partial D)$ such that there exists $\gamma \in C^{\infty}_{(n,n-2)}(\overline{D})$ with $\bar{\partial}\gamma|_{\partial D} = \varphi\omega$. Obviously, $\mathrm{O}_2(\partial D) \subset \mathrm{O}_1(\partial D) \subset \mathrm{O}(\partial D)$.

Let us consider the problem of describing all measures orthogonal under integration on ∂D to the spaces $\mathrm{O}(\partial D)$, $\mathrm{O}_1(\partial D)$ or $\mathrm{O}_2(\partial D)$. In the classical case of $n=1$ and $D=U=\{z_1\colon |z_1|<1\}$, we have $A_0(\partial D) = \mathrm{O}(\partial D)$, so that the theorems which provide solutions to the above problem may be regarded as generalizations of the classical F. and M. Riesz theorem (see, for example, Privalov [1] or Hoffman [1], p. 47):

Let μ be a complex-valued Borel measure on the unit circle ∂U. Then

$$\int_{\partial U} z_1^k \, d\mu = 0, \qquad k = 0,1,2,\ldots, \tag{12.2}$$

if and only if μ is absolutely continuous with respect to Lebesgue measure and there exists $f \in H_1$ such that

$$\mu(E) = \int_E f(z_1)\,dz_1$$

for any Borel set $E \subset \partial U$.

Various methods have been developed which lead to different generalizations of the F. and M. Riesz theorem (see, for example, Glicksberg [1], König and Seever [1], Rainwater [1], Heinz [1], and Khenkin [1]), but they are not applicable in our case, as a rule, since $\mathrm{O}(\partial D)$ is not a ring if $n>1$.

THEOREM 12.1. I. *Let μ be a complex-valued Borel measure on ∂D. Then*

$$\int_{\partial D} \varphi \, d\mu = 0 \tag{12.3}$$

for all $\varphi \in \mathrm{O}(\partial D)$ if and only if there exists $f(z) \in H_1$ such that

$$\mu(E) = \int_E f(z)\,\omega$$

for all Borel subsets E of ∂D.

II. (12.3) *holds for all* $\varphi \in O_1(\partial D)$ *if and only if* μ *is generated by a form* $f\omega$ *with* $f \in H_1$.

III. *If* $n > 1$ *and* ∂D *is connected, then* (12.3) *holds for all* $\varphi \in O_2(\partial D)$ *if and only if* μ *is generated by a form* $f\omega$ *with* $f \in H_1$.

Thus, Theorem 12.1 generalizes the classical F. and M. Riesz theorem. But we may also interpret Theorem 12.1 in another way. Consider the question: under what conditions can a function $f(z)$, defined on ∂D, be extended to the interior of D holomorphically? For $n > 1$, if ∂D is connected, then Hartogs' theorem(2) (Fuks [1], Chapter IV, §21.2, and Shabat [1], p. 349) says that it is sufficient that f be holomorphic in a neighborhood of ∂D. The more general theorem of Severi (Shabat [1], p. 344, or Martinelli [1]) gives the necessary and sufficient condition (3.2) for functions $f(z)$ of class C^1 on ∂D. The local differential conditions may be replaced by global integral ones (§3, 2°). It follows from Theorem 6.1 that, in the case of a strictly pseudoconvex domain D, $\varphi \in O(\partial D)$ if and only if $\varphi\omega$ is $\bar\partial$-exact. In the case of an arbitrary domain D, this condition is at least sufficient to ensure that $\varphi \in O(\partial D)$, as Stokes' theorem immediately shows (regarding necessary and sufficient conditions, see §7). Thus it is natural to call a measure μ a *weak solution of the tangential Cauchy-Riemann equations* if (12.3) holds for all $\varphi \in O(\partial D)$. Then part I of Theorem 12.1 may be reformulated as follows (a similar interpretation may also be given to parts II and III):

THEOREM 12.2. *A complex-valued Borel measure* μ *on* ∂D *is generated by the form* $f\omega$, $f \in H_1$, *if and only if it is a weak solution of the tangential Cauchy-Riemann equations.*

Part I of Theorem 12.1 is a consequence of part II, the proof of which is new even for the classical case $n = 1$. In order to illustrate the method, we carry out the proof for $n = 1$ in subsection 2°, and then indicate the modifications to be made in the case $n > 1$.

2°. For the proof of the Riesz theorem, we shall use the following assertions.

1) *If* $\varphi \in C(\partial U)$ *and*

$$\int_{\partial U} \varphi(z_1) z_1^k dz_1 = 0, \qquad k = 0, 1, \ldots, \tag{12.4}$$

then $\varphi \in AC(\bar U)$.

This proposition can be proved elementarily in various ways. For instance, it follows immediately from the classical theorem of Fejér on the uniform convergence of the sequence of arithmetic means of the partial sums of the

(2) Also referred to sometimes as the Osgood-Brown theorem.

Fourier series of the continuous function φ to that same function. We need only write the Fourier series in the complex form.

Now let X be a Banach space, X^* its dual, and Γ a linear set in X^*. Consider the following closedness properties of Γ:

a) *If* $x_m^* \in \Gamma$ *and* $\lim_{m\to\infty} x_m^*(x) = x^*(x)$ *for all* $x \in X$, *then* $x^* \in \Gamma$.

b) *If* $x_0^* \notin \gamma$, *then there exists an* $x \in X$ *such that* $x_0^*(x) = 1$ *and* $x^*(x) = 0$ *for all* $x^* \in \Gamma$.

2) *For a separable Banach space* X, b) *follows from* a).

For this assertion, we refer to a simple theorem of Banach (Banach [1], Chapter VIII, Theorem 5) asserting that the two conditions are in fact equivalent in the case under consideration.(3)

PROOF OF THE F. AND M. RIESZ THEOREM. Let C^* be the dual space of $C(\partial u)$ and $\Gamma \subset C^*$ the set of measures ν of the form $d\nu = f(z_1)dz_1, f(z_1) \in H_1$. We must show that $\mu \in \Gamma$.

We first show that Γ is closed in the sense of a). Let

$$\begin{gathered} d\nu_m = f_m(z_1)dz_1, \qquad f_m(z_1) \in H_1, \\ \lim_{m\to\infty} \int_{\partial U} \psi(z_1)d\nu_m = \int_{\partial U} \psi(z_1)d\nu_0 \end{gathered} \tag{12.5}$$

for all $\psi(z_1) \in C(\partial U)$. If we apply (12.5) with the Cauchy kernel in the place of ψ and use the fact that a weakly convergent sequence of functionals is bounded in norm, then we see that the sequence $f_m(z_1)$ converges uniformly in the interior of the disc U. The limit function $f_0(z_1)$ obviously belongs to H_1. To prove that $d\nu_0 = f_0(z_1)dz_1$, it suffices to show that

$$\lim_{m\to\infty} \int_{\partial U} \psi(z_1)d\nu_m = \int_{\partial U} \psi(z_1)f_0(z_1)dz_1$$

for all ψ in $C(\partial U)$ or for a set of functions ψ which is dense in $C(\partial U)$. As such a set we take the set of all polynomials $P(z_1, z_1^{-1})$. Each such polynomial can be represented as $P_1(z_1) + P_2(z_1^{-1})$, where $P_1(z_1)$ is then orthogonal to the functions of class H_1 in the sense of integration along ∂U. So we must examine whether

$$\lim_{m\to\infty} \int_{\partial U} P_2\left(\frac{1}{z_1}\right) f_m(z_1)dz_1 = \int_{\partial U} P_2\left(\frac{1}{z_1}\right) f_0(z_1)dz_1. \tag{12.6}$$

The integrals in (12.6) are linear combinations of the values of the respective functions and their derivatives at 0; this is evident for functions in $A(\overline{D})$, and easily follows for H_1-functions by a limiting process. Now (12.6) follows from the uniform convergence of f_m to f_0 inside U.

(3) See also Dunford and Schwartz [1], Chapter V, §7, Exercise 16.

Thus Γ possesses property a), hence also b), by 2). If $\mu \notin \Gamma$, there is a $\varphi \in C(\partial U)$ such that

$$\int_{\partial U} \varphi(z_1) f(z_1) dz_1 = 0 \tag{12.7}$$

for all $f \in H_1$, but

$$\int_{\partial U} \varphi(z_1) d\mu = 1. \tag{12.8}$$

But (12.7) implies (12.4); hence $\varphi(z_1) \in A(\partial U)$ by 1). Any function in $A(\partial U)$ can be approximated by polynomials in z_1, and so (12.8) contradicts (12.2).

LEMMA 12.3. *The formula* (5.5) *is valid also for* $f \in H_1$.

PROOF. We first extend the $w_i(z)$ smoothly to a neighborhood V of ∂D, and then shrink V, if necessary, so that $\langle z, w\rangle \neq 0$ in V. Also we replace w_i by the normalized functions $w_i\langle z, w\rangle^{-1}$, $i = 1, \ldots, n$. Thus we have extended the $w_i(z)$ to a neighborhood of ∂D preserving (5.1). Now we apply (5.5) to the domains D_m, $m > m_0$, and then effect a passage to the limit using properties of H_1-functions.

PROOF OF THEOREM 12.1. The proof proceeds exactly like that of the F. and M. Riesz theorem, using the propeties of H_1-functions, replacing the Cauchy kernel by the Martinelli-Bochner kernel and assertion 1) by Theorems 3.1 and 3.2. Here, instead of the decomposition $P(z_1, z_1^{-1}) = P_1(z_1) + P_2(z_1^{-1})$, we use Corollary 11.5. In the special case of a domain D for which there exists a vector-valued function $w(z)$ such that (5.1) and (5.8) hold and the form ω is nondegenerate on ∂D (fairly large classes of such domains were presented in §5.3°), the proof can be made more elementary by appealing to (5.8) and the decomposition of $P(z, w)$ given by Corollary 5.2 and Lemma 5.3 rather than Corollary 11.5.

§13. On the general form of integral representations of holomorphic functions

1°. The problem of describing forms which are orthogonal to holomorphic functions is closely related to that of describing the general form of integral representation of holomorphic functions. By an *integral representation of holomorphic functions* we mean a formula of the type

$$f(z) = \int_{\partial D} f(\zeta)\mu_z(\zeta), \tag{13.1}$$

valid for all $z \in D$ and all $f \in A(\overline{D})$. Here D is a bounded domain with smooth boundary, and $\mu_z \in C_{2n-1}(\partial D)$ for fixed $z \in D$. A form $\mu_z(\zeta)$ satisfying (13.1) will be called a *reproducing kernel.* A reproducing kernel is a solution of the system inhomogeneous linear equations (13.1), and the orthogonal forms are the solutions of the corresponding system of homogeneous equations. Hence any reproducing kernel can be obtained from a fixed one by the addition of forms orthogonal to holomorphic functions, and conversely, every form which is orthogonal to holomorphic functions can be obtained as the difference between a fixed reproducing kernel and another suitable one.

For functions in $AC(D)$, we have the following well-known integral representation: *if $w(\zeta, z)$ is a vector-valued function which is of class C^1 in the variable $\zeta \in \partial D$ for fixed $z \in D$, and satisfies $\langle w, \zeta - z\rangle = 1$* (throughout this section, $w = (w^1, \ldots, w^n)$ will denote a vector-valued function with the above properties), *then* (*Cauchy-Fantappiè formula*)

$$f(z) = \int_{\partial D} f(\zeta)\omega(w, \zeta, z) \tag{13.2}$$

for all $f \in AC(D)$ ($\omega(w, \zeta, z)$ has been defined in §1.2°).(4)

If we put $w = t = (\bar{\zeta} - \bar{z})/|\zeta - z|^2$, then (13.2) becomes the well-known Martinelli-Bochner formula (see (1.1) for $p = q = 0$).

Observe that, if a reproducing kernel is multiplied by a function $\varphi_z(\zeta)$, $\zeta \in \overline{D}$, $z \in D$, satisfying

$$\varphi_z \in AC(D), \qquad \varphi_z(z) \equiv 1, \qquad z \in D, \tag{13.3}$$

then we again get a reproducing kernel. Indeed, suppose $\mu_z(\zeta)$ satisfies (13.1). Then

$$\int_{\partial D} f(\zeta)\varphi_z(\zeta)\mu_z(\zeta) = f(z)\varphi_z(z) = f(z).$$

It must be noted that the class of kernels of the form $\varphi_z(\zeta)\omega(w, \zeta, z)$ is invariant under biholomorphic maps (Aĭzenberg [5]), though it is not known whether a similar assertion holds with regard to the Cauchy-Fantappiè kernel.

2°. For $n = 1$, the Cauchy-Fantappiè formula goes over into the Cauchy integral formula

$$f(z) = \frac{1}{2\pi i}\int_{\partial D} \frac{f(\zeta)d\zeta}{\zeta - z}.$$

THEOREM 13.1. *Let D be a bounded domain with smooth boundary, and $\mu_z(\zeta)$ a continuous reproducing kernel. Then*

$$\mu_z(\zeta) = \frac{1}{2\pi i} \cdot \frac{\varphi_z(\zeta)d\zeta}{\zeta - z}$$

for some $\varphi_z(\zeta)$ satisfying (13.3).

(4) (13.2) is a consequence of (1.1), Lemma 1.3 and Stokes' theorem.

PROOF. Since $d\zeta$ is nondegenerate on ∂D and continuous, we have

$$\mu_z(\zeta) = g_z(\zeta)d\zeta$$

where $g_z \in C(\partial D)$ for fixed $z \in D$.

Consider the function

$$h_z(\zeta) = g_z(\zeta) - \frac{1}{2\pi i(\zeta - z)}.$$

For every $f \in A(\bar{D})$, we have

$$\int_{\partial D} f(\zeta)h_z(\zeta)d\zeta = 0.$$

By Theorem 3.1, h_z can be extended to D, to each fixed $z \in D$, as a function in $AC(D)$. Set

$$\varphi_z(\zeta) = 1 + 2\pi i(\zeta - z)h_z(\zeta).$$

Then $\varphi_z \in AC(D)$ and $\varphi_z(z) \equiv 1$ for $z \in D$. By the definition of h_z and φ_z, we have

$$g_z(\zeta) = \frac{1}{2\pi i} \cdot \frac{\varphi_z(\zeta)}{\zeta - z}.$$

3°. For strictly pseudoconvex domains in $\mathbf{C}^2$, the Cauchy-Fantappiè formula gives the general form of integral representation, as the next theorem asserts.

THEOREM 13.2. *Let D be a strictly pseudoconvex domain in $\mathbf{C}^2$ with boundary of class C^3, and $\mu_z \in C_3^{1,\lambda}(\partial D)$ a reproducing kernel. Then μ_z is of the form $\omega(w, \zeta, z)$ for some $w \in C^1(\partial D)$.*

PROOF. Every point $\zeta \in \partial D$ has a neighborhood U_ζ such that, on $\partial D \cap U_\zeta$, all differentials can be expressed either in terms of the differentials of z_1, z_2 and $\bar{z}_1$ or those of z_1, z_2 and $\bar{z}_2$. The U_ζ cover ∂D; choose a finite subcovering, say $U_1, \dots, U_m$. Let $\{\varphi_i\}$ be a partition of unity subordinate to the covering $\{U_i\}$. Then $\mu_z = \mu_1 + \cdots + \mu_m$, where $\mu_j = \varphi_j \mu_z$. By the choice of the U_i, we have for $z \in U_i$

$$\mu_i = a_i d\bar{z}_1 \wedge dz + b_i d\bar{z}_2 \wedge dz, \tag{13.4}$$

where a_i, $b_i \in C^{1,\lambda}(U_i)$. But $\operatorname{supp} \mu_i \Subset U_i$; hence $\operatorname{supp} a_i$ and $\operatorname{supp} b_i \Subset U_i$. Therefore, extending a_i and b_i by zero outside U_i, we get $a_i, b_i \in C^{1,\lambda}(\partial D)$, and (13.4) is valid everwhere on ∂D, i.e. $\mu_j \in C_{(2,1)}^{1,\lambda}(\partial D)$. Hence $\mu \in C_{(2,1)}^{1,\lambda}(\partial D)$.

Consider the form

$$\alpha_z(\zeta) = \mu_z(\zeta) - \omega(t(\zeta, z), \zeta, z).$$

It follows from (13.1) and the Martinelli-Bochner formula that $\alpha_z \in A^{\perp}(\bar{D})$. Hence $\alpha_z = \bar{\partial}\gamma_z|_{\partial D}$ by Theorem 6.1, with $\gamma_z \in C^1_{(2,0)}(\bar{D})$. Let g_z be the coefficient of γ_z. Then $g_z \in C^1(\bar{D})$. Set

$$w(\zeta, z) = t(\zeta, z) + (2\pi i)^2 g_z(\zeta)(-(\zeta_2 - z_2), \zeta_1 - z_1).$$

Then $\langle w, \zeta - z\rangle = 1$, and $w \in C^1(\partial D)$. By Lemma 1.6, for $\zeta_1 \neq z_1$ we have

$$\begin{aligned}\omega(w, \zeta, z) &= \frac{1}{(2\pi i)^2} \cdot \frac{\bar{\partial}_\zeta w_2 \wedge d\zeta}{\zeta_1 - z_1} \\ &= \frac{1}{(2\pi i)^2} \cdot \frac{\bar{\partial}_\zeta t_2(\zeta, z) \wedge d\zeta}{\zeta_1 - z_1} + \bar{\partial}_\zeta g_z \wedge d\zeta \\ &= \omega(t(\zeta, z), \zeta, z) + \bar{\partial}\gamma_z = \mu_z(\zeta).\end{aligned}$$

By continuity, the equality $\omega(w, \zeta, z) = \mu_z(\zeta)$ persists for $\zeta_1 = z_1$.

4°. There is a kind of converse to Theorem 13.2:

THEOREM 13.3. *Let D be a domain in $\mathbf{C}^2$ with C^2 boundary, and suppose that every $\mu \in C^\infty_{(2,1)}(\partial D)$ satisfying*

$$f(z^0) = \int_{\partial D} f(\zeta)\mu(\zeta), \qquad f \in A(\bar{D}), \tag{13.5}$$

for some fixed $z^0 \in D$ is of the form $\omega(v(\zeta), \zeta, z^0)$ for some $v \in C^2(\partial D)$ such that $\langle v(\zeta), \zeta - z^0\rangle = 1$. Then D is a domain of holomorphy.

PROOF. We shall show that D satisfies the conditions of Corollary 8.2. Let $\alpha \in A^{\perp}(\bar{D}) \cap C^\infty_{(2,1)}(\partial D)$. Then by assumption there exists $v \in C^2(\partial D)$ such that $\langle v, \zeta - z^0\rangle = 1$ and

$$\omega(t(\zeta, z^0), \zeta, z^0)|_{\partial D} + \alpha = \omega(v, \zeta, z^0), \tag{13.6}$$

since the left side of (13.6) satisfies (13.5). Therefore

$$\alpha = \omega(v, \zeta, z^0) - \omega(t(\zeta, z^0), \zeta, z^0).$$

By Lemma 1.3, the right side is $\bar{\partial}$-exact. An application of Corollary 8.2 now finishes the proof.

5°. For $n > 2$, we have not been able to obtain a description of the general form of integral representation similar to that in Theorem 13.2. We have only the following weaker result.

THEOREM 13.4. *Let D be a strictly pseudoconvex linearly convex domain* (*see Example* 1, §5) *in $\mathbf{C}^n$ with C^3 boundary, and let $\mu_z \in C^{1,\lambda}_{2n-1}(\partial D)$ be a reproducing kernel. Then there exist vector-valued functions $w_1, \ldots, w^{n(n-1)/2} \in C^1(\partial D)$*

such that

$$\mu_z(\zeta) = \sum_{l=1}^{n(n-1)/2} (-1)^l \omega(w^l, \zeta, z) + \lambda_n \omega(t(\zeta, z), \zeta, z), \tag{13.7}$$

where λ_n *is* 1 *if* $n(n-1)/2$ *is even, and* 0 *otherwise.*

PROOF. As in the proof of Theorem 13.2, we get

$$\mu_z(\zeta) = w(t(\zeta, z), \zeta, z)\,|_{\partial D} + \bar\partial\gamma_z\,|_{\partial D}, \tag{13.8}$$

with $\gamma_z \in C^1_{(n,n-2)}(\bar D)$.

Since D is linearly convex, the form $\omega(t(\zeta, z), \zeta, z)$ is nondegenerate on ∂D for fixed $z \in D$ (see Lemma 5.12). Lemma 1.6 implies that for each $\zeta^0 \in \partial D$ there is a j, $1 \leqslant j \leqslant n$, such that $d_\zeta t_1(\zeta, z), \ldots, [j], \ldots, d_\zeta t_n(\zeta, z), d\zeta_1, \ldots, d\zeta_n$ are linearly independent at ζ^0. Hence we have

$$\gamma_z\,|_{\partial D} = \sum_{i \neq j} b_{i,z}(\zeta)\,dt[i, j] \wedge d\zeta$$

in a neighborhood of ζ^0, with $b_{i,z} \in C^1(\partial D)$. As in the proof of (13.4), we deduce that

$$\gamma_z\,|_{\partial D} = \sum_{i<j} a_{i,j,z}(\zeta)\,d_\zeta t[i, j] \wedge d\zeta. \tag{13.9}$$

Observe that $\bar\partial\gamma_z = d\gamma_z$, since γ_z is of type $(n, n-2)$ in ζ. From (13.9) and the invariance of the differential we get

$$\bar\partial\gamma_z\,|_{\partial D} = \sum_{i<j} da_{i,j,z}(\zeta) \wedge d_\zeta t[i, j] \wedge d\zeta. \tag{13.10}$$

Set $\eta_{i,j} = (\zeta_i - z_i)e_j - (\zeta_j - z_j)e_i$, where e_i is the vector with ith coordinate 1 and all others zero. Index the pairs (i, j) with $i < j$ by l, and set

$$w^l = t(\zeta, z) + (-1)^{l+i_l+j_l} \frac{(2\pi i)^n}{(n-1)!} a_{i_l,j_l,z}\eta_{i_l,j_l};$$

then $\langle w^l, \zeta - z\rangle = 1$ for $l = 1, \ldots, n(n-1)/2$, since $\langle \eta_{i,j}, \zeta - z\rangle = 0$. For $\zeta_{i_l} \neq z_{i_l}$ we have

$$\begin{aligned}
\omega(w^l, \zeta, z) &= \frac{(n-1)!}{(2\pi i)^n} \frac{(-1)^{i_l-1} dw^l[i_l] \wedge d\zeta}{\zeta_{i_l} - z_{i_l}} \\
&= \frac{(n-1)!}{(2\pi i)^n} \frac{(-1)^{i_l-1} d_\zeta t[i_l] \wedge d\zeta}{\zeta_{i_l} - z_{i_l}} \\
&\quad + (-1)^{l-1} da_{i_l,j_l,z} \wedge d_\zeta t[i_l, j_l] \wedge d\zeta \\
&= \omega(t(\zeta, z), \zeta, z) + (-1)^{l-1} da_{i_l,j_l,z} d_\zeta t[i_l, j_l] \wedge d\zeta.
\end{aligned}$$

By continuity, this equality also holds for $\zeta_{i_l} = z_{i_l}$. Hence

$$da_{i_l, j_l, z} d_\zeta t[i_l, j_l] \wedge d\zeta = (-1)^l[\omega(t, \zeta, z) - \omega(w^l, \zeta, z)]. \quad (13.11)$$

Summing (13.11) over l from 1 to $n(n-1)/2$, and using (13.8) and (13.9), we get (13.7).

§14. Representation of distributions in $\mathfrak{D}'(\mathbf{R}^{2n-1})$ by $\bar{\partial}$-closed exterior differential forms of type $(n, n-1)$

1°. As is well known (see, for example, Bremermann [1], §5.9), there exists for each distribution T on the real line $\mathbf{R}^1$ a holomorphic function $f(z_1)$ outside the support of T such that T is the difference of the boundary values of $f(z_1)$ on the real axis, in the following sense: for each $\varphi \in \mathfrak{D}$([5])

$$\lim_{\varepsilon \to +0} \int_{\mathbf{R}^1} [f(x + i\varepsilon) - f(x - i\varepsilon)] \varphi(x) dx = \langle T, \varphi \rangle.$$

Such a function f is called an *analytic representation* of T. This concept has turned out to be very useful in the theory of distributions. For example, the product of distributions can be defined with the help of this notion (Bremermann [1], Ivanov [1], Itano [1]).

For distributions on $\mathbf{R}^n$, $n > 1$, analytic representations have also been studied, but have been less useful than in the case $n = 1$, since they consist of 2^n or $n + 1$ holomorphic functions, whose singularities are not contained in the supports of the distributions (see Bremermann [1], §13.8; Martineau [2], Schapira [1], or Vogt [1]).

In this section, we shall define the analytic representation of distributions on $\mathbf{R}^{2n-1}$ by means of $\bar{\partial}$-closed forms of class $B_{n,n-1}$. This representation has the same properties as the analytic representations in the case $n = 1$ (except that holomorphy of the representation is replaced by the property of being $\bar{\partial}$-closed, or more precisely of lying in the class $B_{n,n-1}$). Thus, in this situation also, $\bar{\partial}$-closed forms of type $(n, n-1)$ can be regarded as analogues of holomorphic functions of one variable.

2°. We shall consider the space $\mathbf{R}^{2n-1}$ of the variables $(x_1, \ldots, x_{2n-1})$ and its embedding in the space $\mathbf{C}^n$ of the variables $(z_1, \ldots, z_n)$ that is effected in the following way: $z_j = x_j + ix_{n+j}$, $j = 1, \ldots, n-1$, $z_n = x_n + iy$.

([5]) In §§14 and 15, we use the notation and the properties of the spaces $\mathfrak{D}$, $\mathfrak{D}'$, $\mathfrak{E}$, $\mathfrak{E}'$, $\mathcal{O}_\alpha$ and $\mathcal{O}'_\alpha$ which can be found in Bremermann [1]. In particular, the sign $\langle \ldots, \ldots \rangle$ is used here in a sense different from that of §§1–13.

Observe that the Martinelli-Bochner kernel

$$U_{n,n-1}(\zeta, z) = \frac{(n-1)!}{(2\pi i)^n |\zeta - z|^{2n}} \sum_{k=1}^{n} (-1)^{k-1} \times (\bar{\zeta}_k \times \bar{z}_k) d\bar{z}[k] \wedge dz$$

is a form of class $B_{n,n-1}(\mathbf{C}^n \setminus \zeta)$ in z for fixed ζ.

Let $T \in \mathcal{E}'$. Then the expression

$$\alpha_T(z) = \langle T_\zeta, U_{n,n-1}(\zeta, z) \rangle \tag{14.1}$$

will be called the *Martinelli-Bochner representation* of T. Here $\zeta_j = \eta_j + i\eta_{n+j}$, $j = 1,\ldots,n-1$, and $\zeta_n = \eta_n$, $\eta \in \mathbf{R}^{2n-1}$.

THEOREM 14.1. *If $T \in \mathcal{E}'$ and K is the support of T, then $\alpha_T(z) \in B_{n,n-1}(CK)$, and $\alpha_T = O(|z|^{1-2n})$ as $|z| \to \infty$. The form α_T represents T in the sense that*

$$\lim_{\varepsilon \to +0} \left(\int_{y=\varepsilon} \varphi \alpha_T - \int_{y=-\varepsilon} \varphi \alpha_T \right) = \langle T, \varphi \rangle \tag{14.2}$$

for all $\varphi \in \mathcal{O}_0$. The representation of a derivative of T is the corresponding derivative of α_T.

PROOF. In view of the legitimacy of differentiation under the functional sign, noted earlier in the proof of Theorem 9.1, we have

$$\alpha_T(z) = \left\langle T_\zeta, \partial_z \frac{(n-2)!\,(-1)^n}{(2\pi i)^n |\zeta - z|^{2n-2}} \sum_{k=1}^{n} d\bar{z}[k] \wedge dz[k] \right\rangle$$

$$= \partial_z \sum_{k=1}^{n} \frac{(n-2)!(-1)^n}{(2\pi i)^n} \left\langle T_\zeta, \frac{1}{|\zeta - z|^{2n-2}} \right\rangle d\bar{z}[k] \wedge dz[k],$$

i.e. $\alpha_T(z) \in B_{n,n-1}(\mathbf{C}^n \setminus \{y = 0\})$. Let $z^0 \in \mathbf{R}^{2n-1} \setminus K$. Then there exists $\beta \in \mathfrak{D}$ such that $\beta \equiv 1$ in a neighborhood of K and $\beta \equiv 0$ in a neighborhood of z^0; therefore (see Bremermann [1], §3.11)

$$\langle T_\zeta, U_{n,n-1}(\zeta, z) \rangle = \langle T_\zeta, \beta(\zeta) U_{n,n-1}(\zeta, z) \rangle. \tag{14.3}$$

The right side of (14.3) is also defined in a neighborhood of z^0, and we may differentiate under the "distribution sign". Hence $\alpha_T(z) \in B_{n,n-1}(CK)$. The last assertion of the theorem can be verified similarly.

It remains to prove (14.2). Let us write

$$\alpha_T(z) = \sum_{j=1}^{n} \alpha_j^T(z) d\bar{z}[j] \wedge dz. \tag{14.4}$$

On the plane $\{z: y = \varepsilon\}$, all the summands in (14.4) except the last one vanish, and

$$\frac{(n-1)!\,(-1)^{n-1}}{(2\pi i)^n|\zeta - z|^{2n}}\left[(\bar{\zeta}_n - \bar{z})d\bar{z}[n] \wedge dz - (\bar{\zeta}_n - z_n)d\bar{z}[n] \wedge dz[n] \wedge d\bar{z}_n\right]$$

$$= \frac{(n-1)!|y|\,dx}{\pi^n\left(|\eta - x|^2 + y^2\right)^n} = K(\eta, x, y),$$

where $K(\eta, x, y)$ is the Poisson kernel for the half-space, and $x = (x_1, \ldots, x_{2n-1})$.

Let $T^*(z) = \langle T_\eta, K(\eta, x, y)\rangle$. We shall prove that

$$\lim_{\varepsilon \to +0} \int_{y=\varepsilon} T^*\varphi dx = \langle T, \varphi\rangle \tag{14.5}$$

for all $\varphi \in \mathcal{O}_0$. The integral in (14.5) exists, since $T^*(z) = O(|z|^{-2n})$ as $|z| \to \infty$. Further, the integral can be approximated by its defining sums, which converge in $\mathfrak{S}$ to

$$\left\langle T_\eta, \int_{\mathbf{R}^{2n-1}} \varphi(x)K(x, \eta, \varepsilon)dx\right\rangle = \left\langle T_\eta, \varphi^*(\eta, \varepsilon)\right\rangle,$$

and $\varphi^*(\eta, \varepsilon)$ converges in $\mathfrak{S}$ to $\varphi(\eta)$ as $\varepsilon \to +0$ (see Bremermann [1], §5.4; the lemma in this citation can easily be carried over to the case $n \geqslant 1$). This proves (14.5), and hence (14.2).

Thus the proof of the above theorem proceeds exactly as in the case $n = 1$ (Bremermann [1], §§5.1–5.4, 5.6).

If $T \in \mathfrak{D}'$, then the Martinelli-Bochner representation of T does not exist in general; nevertheless, as in the case $n = 1$, we can formulate

THEOREM 14.2. *Let $T \in \mathfrak{D}'$, and let K be the support of T. Then there is an $\alpha_T \in B_{n,n-1}(CK)$ representing T in the sense that* (14.2) *holds for all $\varphi \in \mathfrak{D}$.*

The proof proceeds as in the case $n = 1$ (Bremermann [1], §5.9). We need only show that, if $T \in \mathfrak{S}'$ and α_T is its Martinelli-Bochner representation, defined in some ball B, then α_T can be approximated uniformly in the interior of B by forms from $B_{n,n-1}(\mathbf{C}^n)$. For this, define

$$\tilde{T}(z) = \left\langle T_\zeta, |\zeta - z|^{2-2n}\right\rangle.$$

Then, as in the proof of Theorem 14.1,

$$\alpha_T(z) = \frac{(-1)^n(n-2)!}{(2\pi i)^n}\partial\left(\tilde{T}(z)\sum_{k=1}^{n} d\bar{z}[k] \wedge dz[k]\right).$$

$\tilde{T}(z)$ is harmonic outside the support of T; hence α_T can be uniformly approximated inside B by forms

$$\frac{(-1)^n(n-2)!}{(2\pi i)^n}\partial\left(P_j(z)\sum_{k=1}^{n} d\bar{z}[k]\wedge dz[k]\right),$$

where the $P_j(z)$ are harmonic polynomials.

The representation (14.1) is also defined for distributions in $\mathcal{O}_0'$, but in this case only the following result, weaker than Theorem 14.1, can be proved.

THEOREM 14.3. *Let $T\in\mathcal{O}_0'$, and let $\alpha_T(z)$ be defined as in* (14.1); *let K be the support of T. Then $\alpha_T\in B_{n,n-1}(CK)$, $\alpha_T(z)\to 0$ as $|z|\to\infty$ and $|\operatorname{Im} z_n|\geqslant c>0$, and* (14.2) *holds for all $\varphi\in\mathcal{O}_{-2n}$.*

PROOF. The proof that $\alpha_T\in B_{n,n-1}(CK)$ is similar to that of the fact that α_T is holomorphic when $n=1$ (Bremermann [1], §6.5). We have

$$(\bar{\zeta}_k-\bar{z}_k)\,|\zeta-z|^{-2n}\to 0\quad\text{in } \mathcal{E} \text{ as } |z|\to\infty$$

and $|\zeta_k-z_k|\cdot|\zeta-z|^{-2n}\leqslant(\operatorname{Im} z_n)^{2-2n}$. We can write down similar inequalities for the derivatives; hence $U_{n,n-1}(\zeta, z)\to 0$ in $\mathcal{O}_0$ as $|z|\to\infty$ and $|\operatorname{Im} z_n|\geqslant c>0$. Hence we may take limits in (14.1). The rest of the proof proceeds as in the case $n=1$ (Bremermann [1], §§5.6, 6.5); the necessary uniform estimates at the corresponding places are valid.

We shall say that a distribution T is *of class A*, if there exists an $\alpha_T\in B_{n,n-1}(\mathbf{C}^n\backslash\mathbf{R}^{2n-1})$ representing T in the sense of Theorem 14.3 and α_n^T (see (14.4)) is bounded on the sets $|\operatorname{Im} z_n|\geqslant c$ for all $c>0$.

Clearly, $\mathcal{O}_0\subset A$.

3°. We give below some examples of distributions and their representations.

1) δ—the Dirac distribution. Then

$$\alpha_\delta(z)=\langle\delta_\zeta, U_{n,n-1}(\zeta,z)\rangle=U_{n,n-1}(0,z)$$
$$=\frac{(n-1)!}{(2\pi i)^n|z|^{2n}}\sum_{k=1}^{n}(-1)^k\bar{z}_k d\bar{z}[k]\wedge dz.$$

2) Define the distribution δ_+ by

$$\langle\delta_+,\varphi\rangle=-\lim_{\varepsilon\to+0}\frac{(n-1)!}{2i\pi^n}\int_{\mathbf{R}^{2n-1}}\frac{\varphi(x)(x_n-i\varepsilon)dx}{(|x|^2+\varepsilon^2)^n},$$

for $\varphi\in\mathfrak{D}$. For $n=1$, this gives the usual δ_+ (Bremermann [1], §7.2). Let us show that $\delta_+\in\mathfrak{D}'$. The expression

$$\frac{(n-1)!\varepsilon}{2\pi^n}\int_{\mathbf{R}^{2n-1}}\frac{\varphi(x)dx}{(|x|^2+\varepsilon^2)^n}$$

differs from the Poisson integral for the half-space only by a constant factor; hence its limit as $\varepsilon \to 0$ exists and equals $\frac{1}{2}\varphi(0)$.

Consider the integral

$$\frac{(n-1)!}{2i\pi^n}\int_{\mathbf{R}^{2n-1}} \frac{x_n\varphi(x)\,dx}{\left(|x|^2+\varepsilon^2\right)^n}.$$

Since $\varphi \in \mathfrak{D}$, we may suppose that the integration extends over a ball $B_r = \{x\colon x \in \mathbf{R}^{2n-1}, |x| < r\}$. Further we replace $\varphi(x)$ by $[\varphi(x) - \varphi(0)] + \varphi(0)$, and observe that

$$\int_{B_r} \frac{x_n\,dx}{\left(|x|^2+\varepsilon^2\right)^n} = 0,$$

since the integrand is odd and the domain of integration is symmetric about 0.

Now

$$\varphi(x) - \varphi(0) = \sum_{j=1}^{2n-1} x_j \int_0^1 \frac{\partial}{\partial x_j}\varphi(tx)\,dt = \sum_{j=1}^{2n-1} x_j\varphi_j(x).$$

The functions $x_k x_n \varphi_k(x)\,|x|^{-2n}$ are integrable in B_r, and

$$|x_k x_n \varphi_k(x)|\left(|x|^2+\varepsilon^2\right)^{-n} \leqslant |x_k x_n \varphi_k(x)|\,|x|^{-2n};$$

hence we may take the limit under the integral sign. Thus δ_+ is defined. It is obviously linear, and its continuity follows from the weak completeness of $\mathfrak{D}'$ (see Shilov [1], Chapter II, §9.3).

Suppose $\varphi \in \mathfrak{D}$ has support outside some ball; then

$$\left|\int_{\mathbf{R}^{2n-1}} \frac{\varphi(x)(x_n - i\varepsilon)\,dx}{\left(|x|^2+\varepsilon^2\right)^n}\right| \leqslant \int_{\mathbf{R}^{2n-1}} \frac{|\varphi(x)|\,|x_n - i\varepsilon|}{|x|^{2n}}\,dx.$$

Taking limits as $\varepsilon \to +0$, we get

$$|\langle \delta_+, \varphi\rangle| \leqslant C\int_{\mathbf{R}^{2n-1}} \frac{|\varphi(x)|\,dx}{|x|^{2n-1}},$$

i.e. $\delta_+ = O(|x|^{1-2n})$. Hence $\delta_+ \in \mathcal{O}'_\alpha$ for any $\alpha < 0$ (see Bremermann [1], §7.2).

Finally, we write down the representation of δ_+:

$$\alpha_{\delta_+}(z) = \begin{cases} \dfrac{(n-1)!}{\left(2\pi i|z|^2\right)^n} \displaystyle\sum_{k=1}^{n} (-1)^k \bar{z}_k\,d\bar{z}[k] \wedge dz, & \operatorname{Im} z_n > 0, \\ 0, \quad \operatorname{Im} z_n < 0. \end{cases}$$

3) $\delta_- = \delta - \delta_+$; hence the representation of δ_- has the form

$$\alpha_{\delta_-}(z) = \begin{cases} 0, & \operatorname{Im} z_n > 0, \\ \dfrac{(n-1)!}{(2\pi i|z|^2)^n} \sum_{k=1}^{n} (-1)^k \bar{z}_k d\bar{z}[k] \wedge dz, & \operatorname{Im} z_n < 0. \end{cases}$$

4) The principal part $P(x_n|x|^{-2n})$ is defined by

$$\langle P(x_n|x|^{-2n}), \varphi \rangle = \lim_{\varepsilon \to +0} \int_{\mathbf{R}^{2n-1}} \frac{x_n \varphi(x)dx}{(|x|^2 + \varepsilon^2)^n}, \qquad \varphi \in \mathfrak{D}.$$

For $n = 1$, this distribution is denoted by $P(x^{-1})$ (Bremermann [1], §7.5). Also,

$$\frac{(n-1)!}{i\pi^n} P\left(\frac{x_n}{|x|^{2n}}\right) = \delta_- - \delta_+ . \tag{14.6}$$

It follows that $P(x_n|x|^{-2n}) \in \mathcal{O}'_\alpha$ for any $\alpha < 0$. As in the case $n = 1$ (Bremermann [1], §7.6), it is easy to show that $P(x_n|x|^{-2n})$ is equivalent to the Cauchy principal value of the integral

$$\int_{\mathbf{R}^{2n-1}} x_n|x|^{-2n} \varphi(x)dx.$$

The representation of this distribution is obvious, taking into account (14.6) and the representations for δ_+ and δ_-.

5) Consider the distributions θ_k, $k = 1, \ldots, n-1$, defined by

$$\langle \theta_k, \varphi \rangle = \lim_{\varepsilon \to +0} \int_{\mathbf{R}^{2n-1}} \frac{(x_n + ix_{n+k})\varphi(x)dx}{(|x|^2 + \varepsilon^2)^n}, \qquad \varphi \in \mathfrak{D}.$$

As in Example 2, it is not hard to show that $\theta_k \in \mathcal{O}'_\alpha$ for any $\alpha < 0$. As a representing form for θ_k, we may take

$$\alpha_{\theta_k}(z) = \begin{cases} (i/2)^{n-1}|z|^{-2n}\left[(-1)^{n-k-1} z_n d\bar{z}[k] \wedge dz + z_k d\bar{z}[n] \wedge dz\right], \\ \qquad\qquad\qquad\qquad\qquad\qquad \operatorname{Im} z_n > 0, \\ 0, \qquad \operatorname{Im} z_n < 0. \end{cases}$$

6) Finally, we define the distribution θ_0 by

$$\langle \theta_0, \varphi \rangle = \int_{\mathbf{R}^{2n-1}} \frac{\varphi(x)dx}{|x|^{2n-2}}.$$

Clearly, $\theta_0 \in \mathcal{O}'_\alpha$ for any $\alpha < -1$, and

$$\alpha_{\theta_0}(z) = \begin{cases} (i/2|z|^2)^{n-1}\big[d\bar{z}[n] \wedge dz + (-1)^{n-k-1} z_n z_k^{-1} \\ \qquad\qquad \times d\bar{z}[k] \wedge dz\big], \quad \operatorname{Im} z_z > 0, \\ 0, \quad \operatorname{Im} z_n < 0, \end{cases}$$

where k can have any integral value from 1 to $n-1$.

It is clear that all the examples are of distributions of class A.

§15. Multiplication of distributions in $\mathfrak{D}'(\mathbf{R}^{2n-1})$

Let $T \in A$. Consider the representation α_T of T which occurs in the definition of the class A. Recall that α_n^T is the coefficient of $d\bar{z}[n] \wedge dz$ in α_T. We denote by B_A the space of all functions of the form

$$\alpha_n^T(z_1, z_2, \ldots, z_n) - \alpha_n^T(z_1, z_2, \ldots, z_{n-1}, \bar{z}_n), \qquad \operatorname{Im} z_n > 0, T \in A.$$

B_A contains differences of all representations of the $T \in A$ satisfying the conditions in the definition of the class A.

The elements of B_A are harmonic functions, bounded on all the sets $\{\operatorname{Im} z_n \geqslant c > 0\}$.

Consider the map ψ: $f \in B_A \to T \in A$ such that for every $\varphi \in \mathfrak{D}$

$$\lim_{\varepsilon \to +0} \int_{y=\varepsilon} \varphi f d\bar{z}[n] \wedge dz = \langle T, \varphi \rangle.$$

We shall show below that ψ is an isomorphism of B_A onto A.

We denote by B_A^* the algebra generated by B_A. We introduce the following equivalence relation in B_A^*: $f_1 \sim f_2$, if, for any $\varphi \in \mathfrak{D}$,

$$\lim_{\varepsilon \to +0} \int_{y=\varepsilon} (f_1 - f_2)\varphi dx = 0.$$

Let $\hat{A}$ be the quotient space of B_A^* by this equivalence relation, and $\hat{f}$ the image of $f \in B_A^*$ under the natural map $B_A^* \to \hat{A}$. We have $B_A \subset \hat{A}$. If

$$T_j \in A \overset{\psi^{-1}}{\to} f_j \in B_A, \qquad j = 1, 2,$$

then by the *product* $T_1 \cdot T_2$ we shall mean the element $(\hat{f}_1 f_2)$ of $\hat{A}$.

We define the *derivative* in $\hat{A}$ by

$$\frac{\partial}{\partial x_k}\hat{f} = (\partial f/\partial x_k)^{\hat{}}.$$

It is defined unambiguously; indeed, if $f_1 \sim f_2$, then

$$\lim_{\varepsilon \to +0} \int_{y=\varepsilon} \left[\frac{\partial}{\partial x_k}(f_1 - f_2)\right] \varphi dx = -\lim_{\varepsilon \to +0} \int_{y=\varepsilon} (f_1 - f_2) \frac{\partial \varphi}{\partial x_k} dx = 0$$

for any $\varphi \in \mathcal{O}_{-2n}$, i.e.

$$\partial f_1/\partial x_k \sim \partial f_2/\partial x_k.$$

THEOREM 15.1. 1) *ψ is an isomorphism of B_A onto A.*

2) *If $S, T \in A$ and the supports of S and T are disjoint, then $S \cdot T = 0$.*

3) *If S and T are defined by bounded continuous functions $S(x)$ and $T(x)$, then $S \cdot T = S(x)T(x)$.*

PROOF. 1) By definition, ψ is surjective, A and B_A are vector spaces, and ψ is linear; hence we need only show that the kernel of ψ is trivial. If $f \in \operatorname{Ker} \psi$, then, for any $\varphi \in \mathcal{D}$,

$$\lim_{\varepsilon \to +0} \int_{\mathbf{R}^{2n-1}} \varphi(x) f(x, \varepsilon)\, dx = 0. \tag{15.1}$$

But since the left side above is a distribution lying in $\mathcal{O}'_{-2n}$ and $\mathcal{D}$ is dense in $\mathcal{O}_{2n}$, (15.1) holds for any $\varphi \in \mathcal{O}_{-2n}$ as well (Bremermann [1], §6.7).

Put

$$\varphi_t(x) = K(x, 't, t_n) = \frac{(n-1)!}{\pi^n} \cdot \frac{t_{2n}}{\left(|x - 't|^2 + t_{2n}^2\right)^n},$$

where $'t = (t_1, \ldots, t_{2n-1})$. The function $f(x, y + \varepsilon)$ is harmonic and bounded for $y \geqslant 0$; hence by Poisson's formula

$$f('t, t_{2n} + \varepsilon) = \int_{\mathbf{R}^{2n-1}} \varphi_t(x) f(x, \varepsilon)\, dx.$$

Letting $\varepsilon \to +0$, we get $f(t) = 0$ for $t_{2n} > 0$.

2) Set $u = \psi^{-1}T$ and $v = \psi^{-1}S$. We must show that, for any $\varphi \in \mathcal{D}$,

$$\lim_{\varepsilon \to +0} \int_{\mathbf{R}^{2n-1}} u(x, \varepsilon) v(x, \varepsilon) \varphi(x)\, dx = 0. \tag{15.2}$$

Outside the support of S, we have $v(x, 0) = 0$; hence $v(x, y) = y v_1(x, y)$ in a neighborhood of any such point. If we now choose φ with support in such a neighborhood, then $\varphi(x) v(x, \varepsilon) \to 0$ in $\mathcal{D}$ as $\varepsilon \to +0$; hence (15.2) holds (see Shilov [1], Chapter II, §9.3). (15.2) also holds for functions with support outside that of T. Since every $\varphi \in \mathcal{D}$ can be represented as the sum of functions φ_j with supports disjoint from that of either S or T (see Bremermann [1], §3.6), it follows that (15.2) also holds in the general case.

3) Let $w = \psi^{-1}(TS)$. Then u, v and w are harmonic extensions of $S(x)$, $T(x)$ and $S(x)T(x)$ to the half-space $\{x\colon \operatorname{Im} z_n > 0\}$, given by the Poisson integral formula. Also, $u \to S$, $v \to T$, and $w \to ST$ uniformly on any compact set (see the proof of Theorem 14.1). Hence $uv \to ST$ uniformly on any compact set, so that $uv \sim w$. Q.E.D.

We remark that, for $n = 1$, several definitions of the product of distributions have been introduced by means of analytic representations (see, for example, Bremermann [1], Russian pp. 245–259)* Ivanov [1], or Itano [1]), but this kind of definitions leads to difficulties for $n > 1$, basically for the reason that the product of distributions with disjoint supports need not be zero; even the product of the zero distribution with a nonzero one need not be zero (Itano [1]).(6) Theorem 15.1 shows that the definition of the product given here does not lead to these difficulties.

In conclusion, we present some examples.

1)

$$\psi^{-1}\theta_k = (i/2)^{n-1} z_k/|z|^{2n}, \quad \text{and} \quad \psi^{-1}\theta_0 = \left(i/2|z|^2\right)^{n-1}.$$

It follows from the identity

$$|z|^{2-2n}(\partial/\partial\bar{z}_k)\left(z_k/|z|^{2n}\right) = -nz_k^2/|z|^{4n}$$

that

$$\theta_0 \circ (\partial/\partial\bar{z}_k)\theta_k = -n\theta_k \circ \theta_k.$$

2)

$$\psi^{-1}\delta = \frac{(n-1)!(-1)^n(z_n - \bar{z}_n)}{\left(2\pi i|z|^2\right)^n},$$

and

$$\psi^{-1}P\left(x_n|x|^{-2n}\right) = \frac{i^{n-1}}{\left(2|z|^2\right)^n}(\bar{z}_n + z_n).$$

We have

$$\theta_0 \circ (\partial/\partial x_n)\delta = -2n\delta \circ P\left(x_n|x|^{-2n}\right), \tag{15.3}$$

since

$$\frac{1}{|z|^{2n-2}}\frac{\partial}{\partial x_n}\frac{z_n - \bar{z}_n}{|z|^{2n}} = -n\frac{z_n - \bar{z}_n}{|z|^{2n}} \cdot \frac{z_n - \bar{z}_n}{|z|^{2n}}.$$

For $n = 1$, (15.3) goes over into the known identity

$$\delta' = -2\delta \circ P(x^{-1})$$

(see Ivanov [1], p. 18).

---

* *Editor's note.* The citation is to an appendix added to the Russian translation.

(6) Other methods of defining the product of distributions have been developed by Vladimirov [2]–[4].

BRIEF HISTORICAL SURVEY AND OPEN PROBLEMS FOR CHAPTERS I–IV

1°. The formula (1.1) was first proved for $(0, q)$-forms by Koppelman [1], and then generalized to (p, q)-forms in Aronov and Dautov [1]. Lemmas 1.1 to 1.4 and 1.6 were formulated by Koppelman [1], [2]. The elementary proof of (1.1) given in §1 is a detailed exposition of the proof in Aronov [1].

The results of §§2, 4, 6, 7 and 13 had been noted previously under stronger smoothness assumptions. The use of Hölder conditions has enabled us to sharpen these assertions in this book.

Theorem 2.5 was formulated in the case $q = n - 1$ by Aĭzenberg [6], [7] and Serbin [1], but the latter's proof is faulty. For arbitrary q, the theorem was proved by Dautov. Theorem 2.6 is due to Dautov and Kytmanov [1]. Corollary 2.7 was observed for $p = 0$ by Lu Qi-keng and Zhong Tong-de [1]. Example 2.9 and Theorem 2.10 were suggested by Dautov. Theorem 2.11 is also due to Dautov; the solvability of this boundary value problem had been established by Andreotti [1] and Andreotti and Hill [1], but their solution was not explicit. Theorem 2.12 is contained in Aĭzenberg [6], [7]. Corollary 2.13 is taken from Aĭzenberg [7].

Theorem 3.1 was proved by Weinstock [1] for domains with smooth boundary, but his proof works only with a stronger smoothness assumption than $\partial D \in C^1$, e.g. if ∂D is a Lyapunov surface. The proof presented here is due to Aronov and Dautov [2], and is based essentially on the same ideas as in the case $n = 1$ (see Muskhelishvili [1], §29.3). Apart from the tangential Cauchy-Riemann equations (3.2), there are other local differential conditions which may be imposed to secure holomorphic extension of $f(z)$ into D (Hörmander [1], Theorem 2.6.13). The possibility of replacing local differential conditions by global integral ones seems to have been first observed by Fichera

[1] (see also Martinelli [1] and Kohn and Rossi [1]). Theorem 3.2 is taken from the work of Weinstock, but the proof presented here is due to Aronov and Dautov [2].

The construction of a "barrier function" (§4.2°) was given by Khenkin [1] (see also Ramírez de Arellano [1] and Øvrelid [1]). Lemma 4.2 was implicitly present in Dautov [1]. This lemma is basic for the proofs of Theorems 4.3, 4.4, 6.1 and 6.4. Theorem 4.3 is equivalent to Theorems 6.1 and 6.4 except for minor details (this form of the assertions of Theorems 6.1 and 6.4 was proposed to the authors by V. P. Palamodov). Theorem 4.4 is due in the C^∞ case to Kohn [1], whose work is based on L^2-estimates for the $\bar{\partial}$-Neumann problem (see also Hörmander [1], Chapter IV). We remark that a solution of the $\bar{\partial}$-Neumann problem by integral formulas, as well as estimates in the uniform metric, were first obtained by Khenkin [2] and Grauert and Lieb [1] (see also Khenkin [3], Lieb [1], Øvrelid [1], Romanov and Khenkin [1], Kerzman [1] and Polyakov [1], [2]).

The results of §5, excluding Theorem 5.6 and Lemma 5.11, are proved in Aĭzenberg [8], [9]. Theorem 5.6 was published in Dautov [1]; for $n = 2$, it has been obtained jointly by Aĭzenberg and Dautov. Lemma 5.11 is due to Kytmanov and Preobrazhenskiĭ [1].

The results of §6 are due to Dautov [1], [2]. An assertion equivalent to Theorem 6.4 in the C^∞ case had been proved earlier by Kohn and Rossi [1]. In Andreotti and Hill [2], there is a local variant of the Hartogs theorem for C^∞ forms, from which one can also deduce a global result (of the type of Theorem 6.4).

Theorem 7.1 is taken from Aĭzenberg [6], [7].

The propositions of §8 are due to Dautov [4].

Theorem 9.1 is a precise analogue (both in formulation and method of proof) of Runge's theorem in one variable (see Hörmander [1], Theorem 1.3.1). The results of §9 and Theorem 10.2 occur in Dautov [3]. The possibility in principle of obtaining such assertions had been observed by Malgrange [2]. The results can also be deduced from Corollary 2.13 and analogous assertions for harmonic functions (see Palamadov [1], p. 377). Corollary 10.3 and Theorem 10.4 are contained in Aĭzenberg's note [7].

The results of §11 are due to Dautov, and were published (except for Theorem 11.4) in [3]. Theorem 11.1 had been proved, in the case when the compact set is the closure of a domain with smooth boundary and the form is C^∞, by Weinstock [2]. For $n = 1$, the similar assertion is due to Browder (see Gamelin [1], Chapter II, Corollary 1.2). Theorem 11.2 for $n = 1$ is a known theorem of Keldysh [1] with a supplementary condition. Theorem 11.4 is a generalization of the theorem of Hartogs and Rosenthal (see Gamelin [1], Chapter II, Theorem 8.5).

The results of §12 were obtained by Aĭzenberg. In less general form, they have appeared in his papers [8] and [9].

The Cauchy-Fantappiè formula (13.2) was proved for convex domains by Leray [1], and for arbitrary domains by Aĭzenberg [10], [5], Koppelman [2], and Kenkin [1]. In the last article, the proof is close to Leray's. The footnote to formula (13.2) in fact Koppelman's proof [2]. Theorem 13.1 was noted by Aĭzenberg [5]; the remaining results of §13 are due to Dautov [14].

The propositions of §§14 and 15 are taken from Aĭzenberg and Kytmanov [1]. The definition of the product of two distributions considered in §15 was first introduced for $n = 1$ by Ivanov [1], [2].

2°. To conclude, we indicate some open problems which we think are of some interest.

1) The Martinelli-Bochner-Koppelman formula is a generalization of the familiar Martinelli-Bochner representation for holomorphic functions to exterior differential forms. Is there a similar generalization of the Cauchy-Fantappiè formula?

2) Can one replace "all $\bar\partial$-closed forms" in Theorem 3.1 by "all the forms $U_{0,0}(\zeta, z)$, $z \in C\bar{D}$"? In other words: Is the representability of a function in a given domain by the Martinelli-Bochner formula sufficient for it to be holomorphic?([1]) This question can be generalized as follows:

3) If $\alpha \in Z^1_{(p,q)}(\bar{D})$, then α is given by (1.1) without the second summand. Conversely, let $\alpha \in C^1_{(p,q)}(\bar{D})$ and suppose it can be represented by (1.1) without a second summand. Does it follow that α is $\bar\partial$-closed in D? The same question can be asked under weaker smoothness conditions (Hölder condition, etc.).

4) Can one assert in Theorem 2.12 that $\gamma_1 \in B_{p,n-1}(D)$ when $0 \leqslant p \leqslant n-1$?

5) Every harmonic function generates a form of class $B_{n,n-1}$ whose coefficients are derivatives of the function. Describe the harmonic functions generating forms lying in $B_{n,n-1}(C\bar{D}) \cap A^{\perp}(\bar{D})$. A simple solution to this problem may enable one to describe the dual of the space of all holomorphic functions in an arbitrary domain $D \subset \mathbf{C}^n$ (or a compact set $K \subset \mathbf{C}^n$) in a simple way, as some quotient space of the space of functions harmonic in the exterior of the domain (or compact set).

6) Does there exist for every domain $D \subset \mathbf{C}^n$ a vector-valued function $w(z)$, $z \in \partial D$, such that $\langle w, z\rangle \equiv 1$ on ∂D and the polynomials $P(w, z)$ are dense in $C(\partial D)$?

([1]) For the ball in $\mathbf{C}^n$, Aronov [2] has answered the question positively.

7) Are the orthogonal polynomials $P(w, z) \in \mathcal{O}(\partial D)$ dense in $\mathcal{O}(\partial D)$? A positive answer to this question would enable us to state the analogue of the Hartogs-Riesz theorem of §12 in a way similar to the classical theorem: one would be considering measures orthogonal to the polynomials in $\mathcal{O}(\partial D)$ described in §5.

8) Elucidate for which domains the second summand in Theorem 7.1 is necessary. The results of §§6–8 solve the problem only in certain cases (the case of strictly pseudoconvex domains from which a finite number of strictly pseudoconvex domains have been removed; the case of domains not covered by Theorem 8.1, when $n = 2$).

9) Characterize the class of domains for which the Cauchy-Fantappiè formula is the general form of the integral representations for holomorphic functions (analogues of the Cauchy integral). The theorems of §13 show that for $n = 2$ this class contains the strictly pseudoconvex domains and consists only of domains of holomorphy. This problem is related to the next two problems.

10) Is the Cauchy-Fantappiè formula invariant under biholomorphic maps?

11) Find the general form of the integral representations of holomorphic functions for an arbitrary bounded domain with smooth boundary.

12) Prove approximation theorems for forms in $Z_{(p,n-1)}$, similar to the well-known results of Vitushkin [1] and Mergelyan [1].

13) Extend the definition of the product of distributions of class A in $\mathbf{R}^{2n-1}$ to arbitrary distributions.

14) The representations of distributions in §14 are closely related to their Fourier transforms when $n = 1$ (see Bremermann [1], Chapter III). Is there a similar convenient connection when $n > 1$? A positive answer to this question would lend further support to our view that forms in $B_{n,n-1}$ (or $Z_{(n,n-1)}$) are good analogues of the holomorphic functions of one variable.

Note. After this book went to press, solutions have emerged for some of the problems listed above. Aronov and Kytmanov [1] have obtained an affirmative answer to question 2) for functions of class $C^1(\overline{D})$. Also, Kytmanov [3] has settled question 3) positively for forms in $C^2_{(p,q)}(\overline{D})$. In another paper [5] by the same author, problem 13) is solved in arbitrary dimension.

CHAPTER V

INTEGRAL PROPERTIES CHARACTERIZING $\bar\partial$-CLOSED DIFFERENTIAL FORMS AND HOLOMORPHIC FUNCTIONS

§16. A characteristic property of $\bar\partial$-closed forms and forms of class B

In this section, we answer question 3) in the "Brief historical survey and open problems" section.

THEOREM 16.1. *Suppose $\partial(D) \in C^{2,\lambda}$ and $\gamma \in C^{1,\lambda}_{(p,q)}(\overline{D})$, $\lambda > 0$. Then γ is a $\bar\partial$-closed in D if and only if, for all $z \in D$,*

$$\gamma(z) = I^1_{p,q}(D,\gamma) - \bar\partial I^2_{p,q-1}(D,\gamma). \tag{16.1}$$

If the coefficients of γ are harmonic in D, then the conditions on ∂D and the behavior of γ on ∂D can be relaxed.

THEOREM 16.2. *Let D be a bounded admissible domain with piecewise smooth boundary, and suppose $\gamma \in C^1_{(p,q)}(D)$ has coeffficients harmonic in D. Then γ is $\bar\partial$-closed in D if and only if* (16.1) *holds.*

Note that, if $q = 0$, then $I^2_{p,q-1}(D,\gamma) = 0$, and $I^1_{p,q}(D,\gamma)$ has coefficients which are harmonic in D. Hence from Theorem 16.2 we deduce

COROLLARY 16.3. *Suppose D is as in Theorem 16.2, and $\gamma \in C^1_{(p,0)}(\overline{D})$. Then, for $z \in D$,*

$$\gamma = I^1_{p,0}(D,\gamma), \tag{16.2}$$

if and only if the coefficients of γ are holomorphic in D.

For the proof, we need to use the double form

$$w_q(\zeta, z) = \frac{(n-1)!}{(2\pi i)^n} \sum_I{}' \sum_{k \notin I} \sigma(I, k) \frac{\bar\zeta_k - \bar z_k}{|\zeta - z|^{2n}} d\bar\zeta\,[I, k] d\bar z_I$$

(a part of $U_{p,q}$), as well as the following lemma.

LEMMA 16.4. *Let D be a bounded admissible domain in $\mathbf{C}^n$, $n > 1$, and suppose $f \in C^1(\bar D)$ is harmonic in D. Then f can be approximated in the topology of $C^1(\bar D)$ by linear combinations of fractions of the form $|\zeta - z^k|^{2-2n}$, $z^k \notin \bar D$.*

PROOF. Since D is admissible and bounded, the space of functions harmonic in a neighborhood of $\bar D$ is dense (in the topology of $C^1(\bar D)$) in the space of functions of class $C^1(\bar D)$ harmonic in D (see Weinstock [3]). Hence we need only prove the lemma for functions harmonic in a neighborhood of $\bar D$.

Now let f be harmonic in the closure of a domain $G \supset \bar D$ with smooth boundary. Then, by Green's formula, for $z \in G$

$$f(z) = \frac{1}{(2n-2)\sigma_{2n}} \int_{\partial G_\zeta} \left[g(\zeta, z) \frac{\partial f(\zeta)}{\partial n} - f(\zeta) \frac{\partial}{\partial n} g(\zeta, z) \right] d\sigma, \quad (16.3)$$

where σ_{2n} is the area of the unit sphere in $\mathbf{C}^n$, and $g(\zeta, z) = |\zeta - z|^{2-2n}$; also, $\partial f(\zeta)/\partial n$ denotes the derivative of f in the direction of the outward normal to ∂G at ζ, and $d\sigma$ the surface element of ∂G. By replacing the integral in (16.3) by the approximating Riemann sums and $\partial g(\zeta, z)/\partial n$ by difference quotients, we obtain the assertion of the lemma.

PROOF OF THEOREMS 16.1 AND 16.2. The necessity follows immediately from the Martinelli-Bochner-Koppelman formula (1.1). Let us prove the sufficiency. If the form

$$\gamma(\zeta) = \sum_I{}' \sum_J{}' \gamma_{I,J}(\zeta) d\bar\zeta_I \wedge d\zeta_J$$

satisfies (16.1), then each of the forms

$$\gamma_J = \sum_I{}' \gamma_{I,J} d\bar\zeta_J$$

also satisfies (16.1), Indeed,

$$I^1_{p,q}(\gamma) = \int_{\partial D} \sum_I{}' \sum_J{}' \gamma_{I,J}(\zeta) d\bar\zeta_I \wedge d\zeta_J \wedge (-1)^{p(n-q-1)} w_q(\zeta, z)$$
$$\wedge \sum_k{}' \sigma(k) d\zeta[k] \wedge dz_k$$
$$= \int_{\partial D} \sum_I{}' \sum_J{}' \left(\gamma_{I,J}(\zeta) d\bar\zeta_I \wedge w_q \wedge d\zeta_J \wedge \sigma(J) d\zeta[J] \wedge dz_J \right)$$

$$= \sum_J{}' \left(\int_{\partial D} \sum_I{}' \gamma_{I,J}(\zeta) d\bar\zeta \wedge w_q \wedge d\zeta \right) \wedge dz_J$$
$$= \sum_J{}' I^1_{0,q}(D, \gamma_J) \wedge dz_J.$$

Similarly,

$$\bar\partial I^2_{p,q-1}(D,\gamma) = \sum_J \bar\partial I^2_{0,q-1}(D,\gamma) \wedge dz_J.$$

Hence it is enough to consider forms of the type $(0, q)$

$$\gamma(z) = \sum_I{}' \gamma_I(z) d\bar z_I.$$

Suppose now that the conditions of Theorem 16.2 are satisfied. If (16.1) holds for γ, then $\gamma_1(z) = I^2_{p,q}(D, \bar\partial\gamma) = 0$ for $z \in D$ by (1.1). But the coefficients of γ_1 are harmonic outside $\bar D$, and tend to 0 as $|z| \to \infty$; also, being convolutions of compactly supported bounded measurable functions with the locally integrable functions $|\zeta|^{-2n}\bar\zeta_k$, they are continuous in $\mathbf{C}^n$. Hence by the uniqueness theorem for harmonic functions we have $\gamma_1 \equiv 0$ on $\mathbf{C}^n$. Writing out this identity, we see that, for every multi-index I $(i_1 < i_2 < \cdots < i_q)$,

$$\int_D \sum_{I_1}{}' \sum_{k \notin I_1} \frac{\partial \gamma_{I_1}}{\partial \bar\zeta_k} d\bar\zeta_k \wedge d\bar\zeta_{I_1} \wedge \sum_{m \notin I} \sigma(I,m) \frac{\partial g}{\partial \zeta_m} d\bar\zeta\,[I,m] \wedge d\zeta = 0. \tag{16.4}$$

By Lemma 16.4, the $\overline{\gamma_I(\zeta)}$ can be approximated in $C^1(\bar D)$ by linear combinations of the $g(\zeta, z^s)$, $z^s \notin \bar D$, $s = 1, 2, \ldots$. Hence (16.4) yields

$$\int_D \sum_{I_1}{}' \sum_{k \notin I_1} \frac{\partial \gamma_{I_1}}{\partial \bar\zeta_k} d\bar\zeta_k \wedge d\bar\zeta_{I_1} \wedge \sum_{m \notin I} \sigma(I,m) \frac{\partial \bar\gamma_I}{\partial \bar\zeta_m} d\bar\zeta\,[I,m] \wedge d\zeta = 0. \tag{16.5}$$

Summing over I in (16.5), we get

$$\int_D |\bar\partial\gamma|^2 d\bar\zeta \wedge d\zeta = 0,$$

i.e. $\bar\partial\gamma = 0$.

Suppose next that we are in the situation of Theorem 16.1. The coefficients of the form $\gamma_2 = I^1_{0,q}(D, \gamma)$ are harmonic in D, and $\gamma_2 \in C^1_{(0,q)}(\bar D)$ (see Corollary 2.4). Further, $\bar\partial\gamma_2 = \bar\partial\gamma$ by (16.1), and hence

$$I^2_{0,q}(D, \bar\partial\gamma_2) = I^2_{0,q}(D, \bar\partial\gamma) = 0.$$

Thus, (16.1) holds for γ_2, and it remains to apply Theorem 16.2.

We have just considered the properties of forms representable by (16.1), i.e. the Martinelli-Bochner-Koppelman formula without the second summand. It is interesting to discuss the class of forms to which the first summand belongs. If we consider (1.1) with $p = n$ and $q = n - 1$, we see that the first summand has the form

$$I^1_{n,n-1}(D,\gamma) = \int_{\partial D} \gamma(\zeta) U_{n,n-1}(\zeta, z).$$

It is clear that this form is of class $B_{n,n-1}(D)$. Forms of this important class have already been encountered before in this book (see Theorems 2.12, 7.1 and 14.1–14.3). Let us present some new properties of this class of forms. They can be written as

$$\gamma = \gamma_G(z) = \sum_{k=1}^{n} (-1)^{k-1} \frac{\partial G}{\partial z_k} d\bar z[k] \wedge dz,$$

with G harmonic in D.

THEOREM 16.5 (UNIQUENESS). *Let $\partial D \in C^1$, and suppose $G \in C^1(\overline{D})$ is harmonic in D. If the restriction of γ_G to ∂D is zero, then $\gamma_G \equiv 0$.*

PROOF.

$$0 = \int_{\partial D} \overline{G}\gamma_G(z) = \int_D d\overline{G} \wedge \gamma_G = \int_D \sum_{k=1}^{n} \left|\frac{\partial G}{\partial z_k}\right|^2 d\bar z \wedge dz.$$

Hence $\partial G/\partial z_k = 0$, $k = 1,\ldots,n$, i.e. G is antiholomorphic in D, and so $\gamma_G \equiv 0$.

We remark that Theorem 16.5 remains valid even for unbounded domains, if we assume $G = O(|z|^{2-2n})$ as $|z| \to \infty$.

THEOREM 16.6. *If $n > 1$ and ∂D is connected, and $\gamma \in C^1_{(n,n-1)}$ is such that*

$$\gamma(z) = I^1_{n,n-1}(D,\gamma), \qquad z \in D, \tag{16.6}$$

then $\gamma \equiv 0$.

PROOF. Consider the functions

$$-\frac{(n-2)!}{(2\pi i)^n} \int_{\partial D} \gamma(\zeta) g(\zeta, z) = \begin{cases} G^+(z), & z \in D, \\ G^-(z), & z \notin \overline{D}. \end{cases}$$

These functions are harmonic (outside ∂D). It follows from the properties of the potential of a simple layer that $G^+ = G^-$ on ∂D. By Theorem 2.5,

$$\gamma(\zeta)\big|_{\partial D} = \gamma_{G^+}\big|_{\partial D} - \gamma_{G^-}\big|_{\partial D}, \tag{16.7}$$

so that (16.6) implies $\gamma_{G^-}|_{\partial D} = 0$. By Theorem 16.5, $\gamma_{G^-} = 0$ outside $\overline{D}$, i.e. G^- is antiholomorphic outside $\overline{D}$ and vanishes at infinity; hence $G^- \equiv 0$. Hence $G^+ \equiv 0$ as well. Now the assertion of the theorem follows from (16.7).

This theorem shows that there are no nontrivial forms representable in D by (16.6) (i.e. by (1.1) with only the first summand). The next result describes forms representable by (16.6) outside D.

THEOREM 16.7. *Let $\partial(D)$ be connected and of class C^1, and suppose $G \in G^1(CD)$ is harmonic in $C\bar{D}$, with $G = O(|z|^{2-2n})$ as $|z| \to \infty$. Then*

$$\gamma_G(z) = -I^1_{n,n-1}(D, \gamma_G), \qquad z \notin \bar{D},$$

if and only if G can be continued antiholomorphically into D.

PROOF. *Necessity*. By considerations similar to those in the proof of Theorem 16.6, we see that G^- is antiholomorphic in D and that $\gamma_G = \gamma_{G^-}$. Hence $G - G^-$ is antiholomorphic in $C\bar{D}$ and tends to zero as $|z| \to \infty$, so that $G = G^-$.

Sufficiency. Suppose G can be continued antiholomorphically into D: denote the continuation by G_1. Then $G_1 \in C^1(\partial D)$ and G_1 is antiholomorphic in D; hence $G_1 \in C^1(\bar{D})$ (Shabat [1], p. 334). Let

$$-\frac{(n-2)!}{(2\pi i)^n}\int_{\partial D}\gamma_G(\zeta)g(\zeta, z) = \begin{cases} G^+(z), & z \in D, \\ G^-(z), & z \notin \bar{D}; \end{cases}$$

then

$$\gamma_{G^+}\big|_{\partial D} - \gamma_{G^-}\big|_{\partial D} = -\gamma_G\big|_{\partial D} = \gamma_{G_1}\big|_{\partial D} - \gamma_G\big|_{\partial D},$$

so that

$$(\gamma_{G^+} - \gamma_{G_1})\big|_{\partial D} = (\gamma_{G^-} - \gamma_G)\big|_{\partial D}. \tag{16.8}$$

Let $D = \{z \in \mathbf{C}^n\colon \rho(z) < 0\}$, where $\rho \in C^1(\mathbf{C}^n)$ and $\operatorname{grad}\rho\big|_{\partial D} \neq 0$. Set

$$\tilde{N}_G = (2i)^n \sum_{k=1}^{n} \frac{\partial G}{\partial z_k}\frac{\partial \rho}{\partial \bar{z}_k}|\operatorname{grad}\rho|^{-1};$$

$\tilde{N}_G$ is defined near ∂D. It is easy to see that

$$\tilde{N}_G\big|_{\partial D}\partial\sigma = \gamma_G\big|_{\partial D},$$

where $\partial\sigma$ is the surface element of ∂D.

Now we see from (16.8) that $N_{G^+ - G_1} = \tilde{N}_{G^- - G}$ for $z \in \partial D$, and $G^+ - G_1 = G^- - G$ on ∂D. It follows from this and the complex form of Green's theorem that the function

$$\Phi(z) = \begin{cases} G^+(z) - G_1(z), & z \in \bar{D}, \\ G^-(z) - G(z), & z \notin \bar{D}, \end{cases}$$

is harmonic in $\mathbf{C}^n$ with $\lim_{|\zeta| \to \infty}\Phi(\zeta) = 0$; hence $\Phi \equiv 0$. Consequently $G^+(z) = G_1(z)$ and $G^-(z) = G(z)$. Hence $\gamma_{G^+} \equiv 0$ and $\gamma_{G^-} = \gamma_G$.

§17. Holomorphy of continuous functions representable by the Martinelli-Bochner integral; criteria for the holomorphy of integrals of the Martinelli-Bochner type

If $p = 0$, the kernel in (16.2) becomes the Martinelli-Bochner kernel

$$U_{0,0}(\zeta, z) = -U_{n,n-1}(z, \zeta) = \frac{(n-1)!}{(2\pi i)^n} \frac{\sum_{k=1}^n (-1)^{k-1} (\bar{\zeta}_k - \bar{z}_k) d\bar{\zeta}\,[k] \wedge d\zeta}{|\zeta - z|^{2n}}$$

and the formula itself reduces to the Martinelli-Bochner formula for holomorphic functions:

$$f(z) = \int_{\partial D_\zeta} f(\zeta) U_{0,0}(\zeta, z), \qquad z \in D, \tag{17.1}$$

which is valid if D is a bounded domain with ∂D piecewise smooth, and $f \in AC(D)$. The Martinelli-Bochner kernel $U_{0,0}(\zeta, z)$ is not holomorphic in z for $n > 1$. Nevertheless, the following sharpening of Corollary 16.3 holds.

THEOREM 17.1. *If D is a bounded domain with piecewise smooth boundary, and the representation* (17.1) *is valid for an $f \in C^1(\bar{D})$, then $f \in A(D)$.*

We may drop the condition $f \in C^1(\bar{D})$, but we must then impose additional conditions on ∂D:

THEOREM 17.2. *If $f \in C(\bar{D})$ is representable in a bounded domain D by the Martinelli-Bochner integral* (17.1), *then f is holomorphic in D in each of the following cases*:

a) $\partial D \in C^2$;

b) $n = 2$, *and ∂D is connected and of class C^1.*

If $f \in AC(D)$, then

$$\int_{\partial D} f(\zeta) U_{0,0}(\zeta, z) = 0, \qquad z \notin \bar{D}. \tag{17.2}$$

As it turns out, (17.2) completely characterizes the traces on ∂D of functions holomorphic in D:

COROLLARY 17.3. *In each of the two cases considered in Theorem* 17.2, *the condition* (17.2) *holds if and only if for $f \in C(\partial D)$ there exists $F \in AC(D)$ such that $F|_{\partial D} = f$.*

Let $P_{k,l}$ be spherical (i.e. homogeneous and harmonic) polynomials forming an orthonormal basis for $L^2(\partial B(0, 1))$, where $\partial B(0, 1)$ is the unit sphere in $\mathbf{C}^n$. Here, l is the degree of the polynomial, and k its index among the polynomials of degree l occurring in the basis, so that $l = 0, 1, 2\ldots$, and $k = 1, \ldots, \sigma(l)$, where

$$\sigma(l) = \frac{2(n+k-1)(2n+k-3)!}{k!\,(2n-2)!}.$$

COROLLARY 17.4. *Suppose D is bounded, and ∂D is connected and of class C^2. Then an $f \in C(\partial D)$ is the trace on ∂D of a function from $AC(D)$ if and only if*

$$\int_{\partial D} f(\zeta) \sum_{j=1}^{n} (-1)^{j-1} \frac{\partial P_{k,l}}{\partial \zeta_j} d\bar{\zeta}\,[j] \wedge d\zeta = 0,$$

$$l = 0, 1, 2, \ldots; \qquad k = 1, 2, \ldots, \sigma(l).$$

For $n = 1$, this corollary is the well-known classical criterion, namely orthogonality to all the monomials ζ^l, $l = 0, 1, 2, \ldots$.

We shall not present the proof of the rather easy Theorem 17.1 here, but refer the reader to Aronov and Kytmanov [1] (see also Corollary 16.3), or to the book by Aĭzenberg and Yuzhakov [1], §1. We proceed to the proof of Theorem 17.2, which is more difficult.

Let $f \in C(\partial D)$, $\partial D \in C^1$; consider the following integral of Martinelli-Bochner type:

$$\int_{\partial D_\zeta} f(\zeta) U_{0,0}(\zeta, z) = \begin{cases} f^+(z), & z \in D, \\ f^-(z), & z \notin \bar{D}. \end{cases}$$

Consider $\lim_{z,z' \to \zeta \in \partial D} [f^+(z) - f^-(z')]$, where z, $z' \to \zeta$ in the same manner as in (2.9). By Theorem 2.6, this limit is equal to $f(\zeta)$, so that Theorem 17.2 and Corollary 17.3 are equivalent. Therefore it suffices to prove Corollary 17.3.

We introduce some notation. Let $z^0 \in \partial D$, $z^+ \in D$ and $z^- \in C\bar{D}$, with the last two lying on the normal to ∂D at z^0 and $|z^+ - z^=| = |z^- - z^0|$. Define

$$\rho(\zeta) = \begin{cases} -\inf\limits_{z \in \partial D} |\zeta - z|, & \zeta \in \bar{D}, \\ \inf\limits_{z \in \partial D} |\zeta - z|, & \zeta \notin \bar{D}, \end{cases}$$

then $D = \{z \in \mathbf{C}^n : \rho(z) < 0\}$.

If $\partial D \in C^2$, then (see Volkov [1], §2) the following assertions hold:

a) There exists a neighborhood U of ∂D such that $\rho \in C^2(U)$.

b) $|\operatorname{grad} \rho| \equiv 1$ in U.

c)

$$\frac{\partial \rho}{\partial z_k}(z^+) = \frac{\partial \rho}{\partial z_k}(z^-) = \frac{\partial \rho}{\partial (z_k)}(z^0), \qquad k = 1, 2, \ldots, n,$$

and $z^\pm \in U$.

Let

$$N_F^{\pm} = \sum_{k=1}^{n} \frac{\partial \rho}{\partial z_k} \frac{\partial F^{\pm}}{\partial \bar z_k}.$$

LEMMA 17.5. *Suppose $\partial D \in C^2$ and $F \in C(\partial D)$; then*

$$\lim_{z^+, z^- \to z^0} (N_F^+(z^+) - N_F^-(z^-)) = 0$$

uniformly in z^0. If N_F^+ can be extended continuously to D, then N_F^- can be continuously extended to CD, and conversely.

PROOF. Let $F \in C(\partial D)$. It is clear that the second part of the lemma follows from the first. Since $N_F^{\pm} = N_{F+C}^{\pm}$ for any constant C, we may assume $F(z^0) = 0$. Observe that

$$d\bar\zeta\,[k] \wedge d\zeta\,|_{\partial D} = (2i)^n (-1)^{k-1} \frac{\partial \rho}{\partial \zeta_k} d\sigma,$$

where $d\sigma$ is the surface element of ∂D. Therefore

$$U_{0,0}(\zeta, z)\,|_{\partial D} = \frac{(n-1)!}{\pi^n} \sum_{k=1}^{n} \frac{\partial \rho}{\partial \bar\zeta_k} \cdot \frac{\bar\zeta_k - \bar z_k}{|\zeta - z|^{2n}} d\sigma,$$

hence

$$N_F^+(2^+) - N_F^-(z^-)$$

$$= -\frac{(n-1)!}{\pi^n} \int_{\partial D} F(\zeta) \sum_{k=1}^{n} \frac{\partial \rho(z^0)}{\partial z_k} \frac{\partial \rho}{\partial \bar\zeta_k} \left[\frac{1}{|\zeta - z^+|^{2n}} - \frac{1}{|\zeta - z^-|^{2n}} \right] d\sigma$$

$$+ \frac{n!}{\pi^n} \int_{\partial D} F(\zeta) \left[\frac{\sum_{k=1}^{n} (\partial \rho(z^0)/\partial z_k)(\zeta_k - z_k^+) \sum_{m=1}^{n} (\partial \rho/\partial \bar\zeta_m)(\bar\zeta_m - \bar z_m^+)}{|\zeta - z^+|^{2n+2}} \right.$$

$$\left. - \frac{\sum_{k=1}^{n} (\partial \rho(z^0)/\partial z_k)(\zeta_k - z_k^-) \sum_{m=1}^{n} (\partial \rho/\partial \bar\zeta_m)(\bar\zeta_m - z_m^-)}{|\zeta - z^-|^{2n+2}} \right] d\sigma.$$

Let us denote the first integral above by I_1, and the second by I_2.

By performing a unitary transformation and a translation, we may take z^0 to 0, and the tangent plane to ∂D at z^0 to the plane $\alpha = \{w \in \mathbf{C}^n\colon \operatorname{Im} w_n = 0\}$. Then ∂D will be defined in a neighborhood of the origin by the system of equations

$$'\zeta = 'w, \qquad \zeta_n = u_n + i\psi(w),$$

where $'\zeta = (\zeta_1, \ldots, \zeta_{n-1})$, $'w = (w_1, \ldots, w_{n-1})$, and $w = ('w, u_n) \in \alpha$. $\psi(w)$ will be of class C^2 in a neighborhood W of 0 in α, and $z^{\pm}$ will be of the form $(0, \ldots, 0, iy_n^{\pm})$.

∂D is a Lyapunov surface with Hölder exponent 1. Hence

$$|\psi(w)| \leqslant c|w|^2, \qquad w \in W,$$

$$\left|\frac{\partial \psi}{\partial x_j}\right| \leqslant c|w|, \quad j = 1,2,\ldots,n, \qquad \left|\frac{\partial \psi}{\partial y_j}\right| \leqslant c|w|, \quad j = 1,2,\ldots,n-1, \tag{17.3}$$

where $z_j = x_j + iy_j$ (see Vladimirov [5], §22.4; these estimates are derived there for $\mathbf{R}^3$, but the proofs easily carry over to $\mathbf{R}^n$, $n \geqslant 3$).

Hence

$$\left|\frac{\partial \rho}{\partial z_k}\right| \leqslant c_1|w|, \qquad \left|\frac{\partial \rho}{\partial \bar{z}_k}\right| \leqslant c_1|w|, \qquad w \in W, \tag{17.4}$$

for $k = 1,\ldots,n-1$, since

$$\frac{\partial \psi}{\partial z_j} = -\frac{\partial \rho}{\partial z_j}\Big/\frac{\partial \rho}{\partial y_n}, \qquad \left|\frac{\partial \rho}{\partial y_n}\right| \leqslant c_2, \qquad w \in W.$$

Here the constants are independent of the point being considered. Furthermore,

$$|\zeta(w)| \leqslant c_3|w|, \qquad w \in W. \tag{17.5}$$

Fix an $\varepsilon > 0$. Choose a ball B in the plane α with center 0, and an $a > 0$, such that

1) $B \subset w$,
2) $|F(\zeta(w))| < \varepsilon$ for $w \in B$,
3) $B \times [-a, a] \subset U$,
4) $c(2|y_n| + c|w|^2) \leqslant d < 1$ for $|y_n| \leqslant a$ and $w \in B$.

For $z = z^{\pm}$, we have $|\zeta(w) - z|^2 = |w|^{+}(y_n - \psi(w))^2$. Hence

$$\frac{1}{|\zeta - z|^2} = \frac{1}{|w - z|^2} \cdot \frac{1}{1 - (2\psi y_n - \psi^2)/|w - z|^2},$$

but

$$\frac{|2\psi y_n - \psi^2|}{|w - z|^2} \leqslant \frac{c|w|^2(2|y_n| + c|w|^2)}{|w|^2 + y_n^2} \leqslant c(2|y_n| + c|w|^2) \leqslant d < 1.$$

Consequently,

$$\frac{1}{1 - (2\psi y_n - gv^2)/|w - z|^2} = \sum_{k=0}^{\infty} \frac{(2\psi y_n - \psi^2)^k}{|\psi - z|^{2k}}$$

$$= 1 + \frac{2\psi y_n - \psi^2}{|w - z|^2} g_1(w, z),$$

where $g_1(w, z)$ is bounded uniformly for $z \in B$ and $|y_n| \leqslant a$. Therefore

$$\frac{1}{|\zeta - z|^{2n}} = \frac{1}{|w - z|^{2n}}\left(1 + \frac{2\psi y_n - \psi^2}{|w - z|^2} g_2\right), \tag{17.6}$$

$$\frac{1}{|\zeta - z|^{2n+2}} = \frac{1}{|w - z|^{2n-2}}\left(1 + \frac{2\psi y_n - \psi^2}{|w - z|^2} g_3\right), \tag{17.7}$$

and g_2 and g_3 are uniformly bounded for $w \in B$ and $|y_n| \leqslant a$.

Let us estimate the integral I_1 over the portion of ∂D corresponding to the ball, i.e. the integral I_1^B. By (17.6) and (17.3), we have

$$\left|\frac{1}{|\zeta - z^+|^{2n}} - \frac{1}{|\zeta - z^-|^{2n}}\right| \leqslant 2\frac{|2\psi y_n^+ - \psi^2|}{|w - z^+|^{2n+2}}|g_2| \leqslant c_4 \frac{2|y_n^+| + c|w|^2}{\left(|w|^2 + y_n^{+2}\right)^n}.$$

Further, $d\sigma \leqslant d_1\, dS$, where dS is the surface element of the plane α; hence

$$\begin{aligned}|I_1^B| &\leqslant \varepsilon c_5 \int_B \frac{2|y_n^+| + c|w|^2}{\left(|w|^2 + y_n^{+2}\right)^n} dS \\ &\leqslant \varepsilon c_5\left[2\int_{\mathbf{R}^{2n-1}} \frac{|y_n^+|\, dS}{\left(|w|^2 + y_n^{+2}\right)^n} + c\int_B \frac{dS}{\left(|w|^2 + y_n^{+2}\right)^{n-1}}\right].\end{aligned}$$

The first integral above is a Poisson integral for a half-space (see, for example, Stein [2], Chapter III, §2.1) and is independent of y_n^+, while the second integral is continuous at 0. Hence $|I_1^B| \leqslant m\varepsilon$, where m is independent of F and the point considered.

The integral I_1 over the remaining portion of ∂D can be obviously made arbitrarily small as $z^{\pm} \to 0$.

It is easy to see that the form of I_2 is unchanged by the unitary transformation and translation. Hence

$$\begin{aligned}&\sum_{k=1}^{n} \frac{\partial\rho(0)}{\partial z_k}(\zeta_k - z_k) \sum_{m=1}^{n} \frac{\partial\rho}{\partial\bar{\zeta}_m}(\bar{\zeta}_m - \bar{z}_m) \\ &\quad = i(\zeta_n - z_n)\sum_{m=1}^{n} \frac{\partial\rho}{\partial\bar{\zeta}_m}(\bar{\zeta}_m - \bar{z}_m) \\ &\quad = i\sum_{m=1}^{n-1} \frac{\partial\rho}{\partial\bar{\zeta}_m}\bar{\zeta}_m(\zeta_n - z_n) + i\frac{\partial\rho}{\partial\bar{\zeta}_n}\left(u_n^2 + \psi^2 + y_n^2 - 2\psi y_n\right).\end{aligned}$$

The integral I_2 over the part of ∂D corresponding to B can be split into three integrals

$$I_2' = -\frac{2in!}{\pi^n}\int_B F(\zeta(w))\frac{(\partial\rho/\partial\bar\zeta_n)2\psi y_n^+ + \Sigma_{m=1}^{n-1}(\partial\rho/\partial\bar\zeta_m)\bar\zeta_m y_n^+}{|w-z|^{2n+2}}\,d\sigma',$$

$$I_2^{\pm} = \pm\frac{in!}{\pi^n}\int_B F(\zeta(w))$$

$$\times\frac{(\partial\rho/\partial\bar\zeta_n)(u_n^2+\psi^2+y_n^{\pm 2}-2\psi y_n^{\pm}) + \Sigma_{m=1}^{n-1}(\partial\rho/\partial\bar\zeta_m)\bar\zeta_m(\zeta_n - z_n^{\pm})}{|w-z^+|^{2n+4}}$$

$$\times(2\psi y_n^{\pm}-\psi^2)g_3\,d\sigma^1,\qquad ds^1 = \sqrt{1+\sum_{j=1}^{n}\left(\frac{\partial\psi}{\partial x_j}\right)^2+\sum_{j=1}^{n-1}\left(\frac{\partial\psi}{\partial y_j}\right)^2}\,dS,$$

and each of these can be estimated in the same way as I_1^B, using (17.3)–(17.5) and (17.7).

LEMMA 17.6. *Let $D\subset \mathbf{C}^n$ be a domain with C^1 boundary, and let $F\in C(\partial D)$. Set*

$$\int_{\partial D} F d_\zeta g\wedge d\bar\zeta\,[k,m]\wedge d\zeta = \begin{cases} H_F^+(z), & z\in D,\\ H_F^-(z), & z\notin \bar D,\end{cases}$$

where $g(\zeta,z)=|\zeta-z|^{2-2n}$. Then

$$\lim_{z^+,\,z^-\to z^0}(H_F^+(z^+)-H_F^-(z^-))=0,$$

uniformly in z^0. If $H_F^+(z)$ can be extended continuously to $\bar D$, then $H_F^-(z)$ can be extended continuously to CD, and conversely ($z^{\pm}$ and z^0 are as in Lemma 17.5).

PROOF. We essentially repeat the proof of Theorem 2.6. Since $H_F^{\pm}=H_{F+C}^{\pm}$, we may suppose $F(z^0)=0$. As in Lemma 17.5, we make a unitary transformation and translation so that ∂D may be defined in a neighborhood of the origin by the system of equations

$${}'\zeta = {}'w,\qquad \zeta_n = U_n + i\psi(w)$$

(we preserve the notation of Lemma 17.5). In addition, $\psi(w)$ is of class C^1 in a neighborhood w of 0 in the plane α, and $\psi(w)=O(|w|)$.

Fix an $\varepsilon>0$ and choose a ball B in α with center 0 such that

1) $B\subset w$,
2) $|F(\zeta(w))|<\varepsilon$ for $w\in B$,
3) $|w-z^{\pm}|\leqslant c\,|\zeta(w)-z^{\pm}|$ for $w\in B$.

Condition 3) is guaranteed by the relations

$$|w-\zeta(w)|=|\psi(w)|=O(|w|),$$
$$|w-z^{\pm}|\leqslant|w-\zeta(w)|+|\zeta(w)-z^{\pm}|=|\psi(w)|+|\zeta-z^{\pm}|.$$

Let $B'=\{\zeta\in\partial D\colon \zeta=\zeta(w),\ w\in B\}$. Because of 3) above and the inequality $|\zeta(w)|\leqslant C_1|w|\leqslant C_1|w-z^{\pm}|$, we have

$$\left|\frac{\bar\zeta_k}{|\zeta-z^+|^{2n}}-\frac{\bar\zeta_k}{|\zeta-z^-|^{2n}}\right|$$
$$=\frac{||\zeta-z^+|-|\zeta-z^-||}{|\zeta-z^+||\zeta-z^-|}\sum_{j=0}^{2n-1}\frac{|\phi_k|}{|\zeta-z^+|^{j}|\zeta-z^-|^{2n-j-1}}$$
$$\leqslant C_1C^{2n+1}\frac{2(2n-1)|z^+|}{|w-z^+|^{2n}}. \tag{17.8}$$

Similarly,

$$\left|\frac{z_k^+}{|\zeta-z^+|^{2n}}-\frac{z_k^-}{|\zeta-z^-|^{2n}}\right|\leqslant d\frac{|z^+|}{|w-z^+|^{2n}}. \tag{17.9}$$

Finally,

$$d\sigma\leqslant d_1\,dS. \tag{17.10}$$

Now condition 2) on B and (17.8)–(17.10) yield

$$\left|\int_{B'}F(\zeta)\left[d_\zeta g(\zeta,z^+)\wedge d\bar\zeta\,[k,m]\wedge d\zeta-d_\zeta g(\zeta,z^-)\wedge d\bar\zeta\,[k,m]\wedge d\zeta\right]\right|$$
$$\leqslant \varepsilon d_2\int_B\frac{|y_n^+|\,ds}{\left(|w|^2+y_n^{+2}\right)^n}\leqslant\varepsilon d_3.$$

Here d_3 is independent of z^0 and F. Thus the first part of the lemma is proved, and the second part follows from the first, as in Lemma 17.5.

PROOF OF COROLLARY 17.3. Suppose f as in Corollary 17.3, i.e. $f^-\equiv 0$. Then, by the jump formula for integrals of Martinelli-Bochner type (Theorem 2.6), we see that f^+ can be extended continuously to $\bar D$ and its restriction to ∂D coincides with f. Consider now a function φ harmonic in some neighborhood of $\bar D$. There exists a domain G with smooth boundary such that $\bar D\subset G$ and φ is harmonic on $\bar G$. Then, by Green's formula (16.3),

$$\varphi(z)=\frac{(n-2)!}{(2\pi i)^n}\int_{\partial G_\zeta}\left[g(\zeta,z)\frac{\partial\varphi(\zeta)}{\partial n}-\varphi(\zeta)\frac{\partial g(\zeta,z)}{\partial n}\right]d\sigma,\qquad z\in G.$$

Replacing the integral on the right by approximating Riemann sums, and the derivatives by difference quotients, we see that φ can be approximated uniformly in some neighborhood $U(\bar{D})$ of D by linear combinations of fractions of the form $g(\zeta, z^k)$, $z^k \notin U(\bar{D})$, $k = 1, 2, \ldots$. Since

$$U_{0,0}(\zeta, z) = -\frac{(n-2)!}{(2\pi i)^n} \sum_{k=1}^{n} (-1)^{k-1} \frac{\partial g}{\partial \zeta_k} d\bar{\zeta}\,[k] \wedge d\zeta,$$

it follows from (17.2) that

$$\int_{\partial D} f(\zeta) \sum_{k=1}^{n} (-1)^{k-1} \frac{\partial \varphi}{\partial \zeta_k} d\bar{\zeta}\,[k] \wedge d\zeta = 0 \tag{17.11}$$

for all φ harmonic in a neighborhood of $\bar{D}$.

Consider now a sequence of domains D_m, $m = 1, 2, \ldots$, with smooth boundaries, such that $\bar{D}_m \subset D_{m+1}$, $\bigcup_1^\infty D_m = D$, and

$$\lim_{m\to\infty} \int_{\partial D_m} \alpha = \int_{\partial D} \alpha \tag{17.12}$$

for all continuous $(2n-1)$-forms α on $\bar{D}$.

By Stokes' formula

$$(-1)^n \int_{\partial D_m} f^+ \sum_{k=1}^{n} (-1)^{k-1} \frac{\partial \varphi}{\partial \zeta_k} d\bar{\zeta}\,[k] \wedge d\zeta = \int_{\partial D_m} \varphi \mu_{f^+},$$

where $\mu_{f^+} = \sum_{k=1}^{\infty} (-1)^{k-1} (\partial f^+/\partial \bar{\zeta}_k) d\zeta[k] \wedge d\bar{\zeta}$. The integral on the left tends to zero by virtue of (17.11) and (17.12), so that

$$\lim_{m\to\infty} \int_{\partial D_m} \varphi \mu_{f^+} = 0 \tag{17.13}$$

for all functions φ harmonic in a neighborhood of $\bar{D}$.

We now want to show that (17.13) holds for all φ continuous on $\bar{D}$. Using Lemma 17.5 we finish the proof of case a) of Corollary 17.3.

Choose $D_m = \{z \in \mathbf{C}^n \colon \rho(z) < -1/m\}$. Then clearly (17.12) holds. It is easy to verify that

$$d\zeta[k] \wedge d\bar{\zeta}\,|_{\partial D_m} = (2i)^n (-1)^{n+k-1} \frac{\partial \rho}{\partial \zeta_k} d\sigma_m,$$

where $d\sigma_m$ is the surface element of ∂D_m. Hence

$$\mu_{f^+}|_{\partial D_m} = (-2i)^n N_f^+ \, d\sigma_m.$$

Since $N_f^- \equiv 0$, N_f^+ can be extended continuously to $\bar{D}$ by Lemma 17.5, and vanishes on ∂D. Therefore (17.13) holds for all φ in $C(\bar{D})$. Choosing $\bar{f}^+$ as φ, we then get by Stokes' theorem

$$\int_{\partial D_m} \bar{f}^+ \mu_{f^+} = (-2i)^n \int_{\partial D_m} \sum_{k=1}^{n} \left| \frac{\partial f^+}{\partial \bar{\zeta}_k} \right|^2 dV \to 0$$

as $m \to \infty$. Hence f^+ is holomorphic in D.

We have already shown that the restriction of f^+ to ∂D coincides with f.

We now consider case b) of Corollary 17.3. Let $n = 2$, and suppose $\partial(D)$ is connected and of class C^1. We have proved that if $f \in C(\bar{D})$ satisfies (17.2), then (17.11) holds for f. Consider the function

$$\varphi_z(\zeta) = \frac{a(\zeta_1 - z_1) + b(\zeta_2 - z_2)}{b(\bar{\zeta}_1 - \bar{z}_1) - a(\bar{\zeta}_2 - \bar{\zeta}_2)} \cdot \frac{1}{|\zeta - z|^2}.$$

This is harmonic in ζ in $D_1 = \mathbf{C}^2 \backslash \{\zeta\colon b(\bar{\zeta}_1 - \bar{z}_1) - a(\bar{\zeta}_2 - \bar{z}_2) = 0\}$. Let V be a ball containing $\bar{D}$. Then, for each $z \in \bar{V}$, we can choose a and b such that $\bar{V} \subset D_1$. As is easily verified,

$$dg \wedge d\zeta = \left(\frac{\partial \varphi_z}{\partial \zeta_1} d\bar{\zeta}_2 - \frac{\partial \varphi_z}{\partial \zeta_2} d\bar{\zeta}_1 \right) \wedge d\zeta.$$

Thus, by (17.11),

$$\int_{\partial D} f dg \wedge d\zeta = 0, \qquad z \in \bar{V}.$$

But since ∂D is connected, the above integral vanishes for all $z \in C\bar{D}$.

We can now conclude the proof of Corollary 17.3 with the aid of Lemma 17.6. Indeed, we have shown that $H_f^- \equiv 0$; therefore, by Lemma 17.6, H_f^+ can be extended continuously to $\bar{D}$ and vanishes on ∂D. Being harmonic in D, H_f^+ therefore vanishes identically in D.

Further,

$$\frac{\partial}{\partial \bar{z}_1} U_{0,0}(\zeta, z) = \frac{1}{4\pi^2} \frac{\partial}{\partial z_2} dg \wedge d\zeta,$$

$$\frac{\partial}{\partial \bar{z}_2} U_{0,0}(\zeta, z) = -\frac{1}{4\pi^2} \frac{\partial}{\partial z_1} dg \wedge d\zeta.$$

Thus

$$\frac{\partial}{\partial \bar{z}_j} f^+(z) = \pm \frac{1}{4\pi^2} \int_{\partial D} f(\zeta) \frac{\partial}{\partial z_{3-j}} dg \wedge d\zeta = 0, \qquad z \in D, j = 1, 2,$$

i.e. f^+ is holomorphic in D.

PROOF OF COROLLARY 17.4. Since f^- is harmonic in $C\bar{D}$ (and $C\bar{D}$ is connected), we will have $f^-(z) = 0$ for all $z \in C\bar{D}$ if we show that $f^-(z) = 0$

outside a ball B containing $\bar{D}$. For $z \notin B$, $g(\zeta, z)$ is harmonic in B with respect to ζ, and hence can be represented as a series in the basis of polynomials $P_{k,l}(\zeta)$, converging uniformly in B. Hence the equality in Corollary 17.4 implies (17.2). Conversely, if f extends holomorphically into D as a function of class $AC(D)$, then the above-mentioned equality follows from the fact that the integrands are $\bar{\partial}$-closed forms.

To obtain another application of Corollary 17.3, we need the following notion. Suppose D is a bounded domain with $\partial D \in C^1$, and let $f \in C(\partial D)$. We shall say that f is *holomorphically continuable in dimension one* if, for each complex line l, corresponding to the restriction $f_l = f|_{\partial D \cap l}$ there exists $F_l \in C(\bar{D} \cap l)$, holomorphic at interior (relative to l) points of $\bar{D} \cap l$, such that $f = F_l$ on $\partial D \cap l$.

COROLLARY 17.7. *If D is bounded and $\partial D \in C^2$, and if $f \in C(\partial D)$ is holomorphically continuable in dimension one, then f is holomorphically continuable into D.*

This is a consequence of Corollary 17.3 and the following lemma:

LEMMA 17.8. *If a function is holomorphically continuable in dimension one, then* (17.2) *holds for $z \in C\bar{D}$.*

PROOF. Suppose z lies in the unbounded component of $C\bar{D}$. By performing an orthogonal linear transformation and translation in $\mathbf{C}^n$, we may ensure that $z = 0$ and that the plane $\{z\colon z_1 = 0\}$ does not meet D. This can be done for z far enough from D. Thus $U_{0,0}(\zeta, z)$ becomes $U_{0,0}(\zeta, 0)$. We now make the change of variables $\zeta_1 = v_1^{-1}$, $\zeta_j = v_j v_1^{-1}$ for $j = 2, \ldots$. Then

$$d\zeta = d\zeta_1 \wedge \cdots \wedge d\zeta_n = \begin{vmatrix} -\dfrac{1}{v_1^2} & 0 & \cdots & 0 \\ -\dfrac{v_2}{v_1^2} & \dfrac{1}{v_1} & \cdots & 0 \\ \vdots & \vdots & \vdots & \vdots \\ -\dfrac{v_n}{v_1^2} & 0 & \cdots & \dfrac{1}{v_1} \end{vmatrix} dv = -\frac{1}{v_1^{n+1}}\, dv.$$

Also, $\mathbf{C}^n \setminus \{\zeta\colon \zeta_1 = 0\}$ is mapped biholomorphically onto $\mathbf{C}^n \setminus \{v\colon v_1 = 0\}$. Observe that

$$\sum_{k=1}^{n} (-1)^{k-1} \bar{\zeta}_k d\bar{\zeta}\,[k] = \sum_{k=1}^{n} \bar{\Delta}_k d\bar{v}[k]$$

where

$$\Delta_k = \begin{vmatrix} \zeta_1 & \zeta_2 & \cdots & \zeta_n \\ \zeta_1' v_1 & \zeta_2' v_1 & \cdots & \zeta_n' v_1 \\ & \cdots & & \\ & [k] & & \\ & \cdots & & \\ \zeta_1' v_n & \zeta_2' v_n & \cdots & \zeta_n' v_n \end{vmatrix} = \begin{cases} \dfrac{1}{v_1^n}, & k = 1, \\ 0, & k > 1. \end{cases}$$

Hence

$$U_{0,0}(\zeta, 0) = (-1)^n \frac{dv_1}{v_1} \wedge \frac{d\bar v[1] \wedge dv[1]}{(1 + |v_2|^2 + \cdots + |v_n|^2)^n} = \frac{dv_1}{v_1} \wedge \lambda('v).$$

Here, $'v = (v_2, \ldots, v_n)$. Under our change of variables, the lines $\{\zeta\colon \zeta_1 = t, \zeta_2 = c_2 t, \ldots, \zeta_n = c_n t\}$ go over to the lines $\{v\colon v_2 = c_2, \ldots, v_n = c_n\}$, and the boundary of D to a certain cycle Γ. If f is holomorphically continuable in dimension one, then, after the change of variables, the new function $f(v)$ will be continuable in dimension one along all the lines $\{v\colon v_2 = c_2, \ldots, v_n = c_n\}$. Furthermore,

$$\int_{\partial D} f(\zeta) U_{0,0}(\zeta, 0) = \int_\Gamma f(v) \frac{dv_1}{v_1} \wedge \lambda('v) = \int_{\pi(\Gamma)} \lambda('v) \int_{\pi^{-1}('v) \cap \Gamma} f(v) \frac{dv_1}{v_1}$$

where π is the projection $v \to {}'v$. Note that, by Sard's theorem, the $\pi^{-1}('v) \cap \Gamma$ are smooth curves for almost all $'v \in \pi(\Gamma)$. The origin lies outside the domain bounded by Γ; hence the inside integral above vanishes for almost all $'v$, i.e. (17.2) holds at points z sufficiently far from D. Since an integral of the Martinelli-Bochner type is harmonic, it follows that (17.2) holds on the unbounded component of $C\bar D$.

Suppose now that z belongs to a bounded component Q of $C\bar D$, and suppose $z = 0$ (we can always achieve this by performing a translation). Consider a cone Π_ε with vertex 0, formed by complex lines through 0, where ε denotes the area of the portion of ∂Q lying inside Π_ε. By the same change of variables as before, this cone goes over into the cylinder formed by the lines $\{v\colon v_2 = c_2, \ldots, v_n = c_n\}$. And Π_ε can be chosen so that: 1) as $\varepsilon \to 0$, the area of $\partial(Q \cap \Pi_\varepsilon) \cap \partial D$ tends to 0, therefore

$$\int_{\partial D} f(\zeta) U_{0,0}(\zeta, 0) = \lim_{\varepsilon \to 0} \int_{\partial D \setminus \partial(Q \cap \Pi_\varepsilon)} f(\zeta) U_{0,0}(\zeta, 0),$$

and 2) $\partial D \setminus \partial(Q \cap \Pi_\varepsilon)$ maps to a compact set Γ_ε under the change of variables. Then it can be shown, as before, that

$$\int_{\Gamma_\varepsilon} f(v) \frac{dv_1}{v_1} \wedge \lambda('v) = 0.$$

Hence (17.2) also holds for $z \in Q$.

Corollary 17.3 is closely related to the criterion for the holomorphy of integrals of Martinelli-Bochner type.

THEOREM 17.9. *Suppose $n > 1$, and D is bounded with ∂D connected and of class C^1; and let $F \in C^1(\partial D)$. If F^+ is holomorphic in D, then F^+ extends continuously to $\bar{D}$ and its restriction to ∂D coincides with F.*

THEOREM 17.10. *Suppose $n > 1$, and D is a bounded domain with ∂D connected and of class C^2. Let $F \in C(\partial D)$. If F^+ is holomorphic in D and extends continuously to $\bar{D}$, then its restriction to ∂D coincides with F.*

The content of the above two theorems is that, in contrast to the one-dimensional case (integrals of Cauchy type), an integral of Martinelli-Bochner type is holomorphic in the case $n > 1$ if and only if it is a Martinelli-Bochner integral, i.e. $F^+ = F$ on ∂D (or equivalently, $F^- \equiv 0$). The condition that ∂D be connected cannot be dropped, as the example of the spherical shell $D = \{z: 1 < |z| < 2\}$ and the function

$$F(\zeta) = \begin{cases} 1, & |\zeta| = 2, \\ 0, & |\zeta| = 1 \end{cases}$$

shows. In this case, $F \in C^1(\partial D)$, but $F^+ \equiv 1$, i.e. F^+ does not coincide with F on $\partial B(0, 1) \subset \partial D$.

For the proof of Theorems 17.9 and 17.10, we need the following assertion on the saltus of derivatives of an integral of Martinelli-Bochner type (Aronov [3]).

LEMMA 17.11. *Let D be a bounded domain with $\partial D \in C^1$: $D = \{z \in \mathbf{C}^n: \rho(z) < 0\}$, where $\rho \in C^1(\mathbf{C}^n)$ and $\operatorname{grad} \rho\,|_{\partial D} \neq 0$. If $F \in C^1(\partial D)$, then*

$$\lim_{z^+ \to z^0} \left(\frac{\partial F^+(z^+)}{\partial \bar{z}_k} - \frac{\partial F^-(z^-)}{\partial \bar{z}_k} \right) = \frac{\partial F(z^0)}{\partial \bar{z}_k} - \rho_{\bar{k}} \sum_{m=1}^{n} \frac{\partial F(z^0)}{\partial \bar{z}_m} \rho_m,$$

uniformly in z^0. Here $z^0 \in \partial D$, $z^\pm$ lie on the normal to ∂D at z^0, $z^+ \in D$, $z^- \notin \bar{D}$, $|z^+ - z^0| = |z^- - z^0|$, and

$$\rho_k(z^0) = \frac{\partial \rho(z_0)}{\partial z_k} |\operatorname{grad} \rho(z^0)|^{-1}, \qquad \rho_{\bar{k}} = \bar{\rho}_k.$$

PROOF OF THEOREM 17.9. We extend F to a function $\tilde{F}$ of class C^1 in a neighborhood of $\bar{D}$. For $\tilde{F}$, the Martinelli-Bochner formula for smooth functions (the special case $p = q = 0$ of (1.1)) is valid:

$$\tilde{F}(z) = \int_{\partial D} F(\zeta)U_{0,0}(\zeta, z) - \int_{D} \bar{\partial}\tilde{F}(\zeta) \wedge U_{0,0}(\zeta, z)$$

$$- \int_{D} \bar{\partial}\tilde{F}(\zeta) \wedge U_{0,0}(\zeta, z). \qquad (17.14)$$

The second integral in (17.14) is the convolution of a bounded compactly supported measurable function with the locally integrable functions $|\bar{\zeta}_k - \bar{z}_k| \cdot |\zeta - z|^{-2n}$, and hence is continuous in $\mathbf{C}^n$. Consequently the first integral, i.e. F^+, extends continuously to $\bar{D}$. Now F^+ is holomorphic in D, and therefore $\partial F^+/\partial \bar{z}_k = 0$; hence it follows from Lemma 17.11 that the $\partial F^-/\partial \bar{z}_k$ extend continuously to CD and their restriction to ∂D has the form

$$\frac{\partial F^-}{\partial \bar{z}_k} = \frac{\partial F}{\partial \bar{z}_k} - \rho_{\bar{k}} \sum_{m=1}^{n} \frac{\partial F}{\partial z_m}\rho_m, \qquad k = 1, 2, \ldots, n. \qquad (17.15)$$

Multiplying (17.15) by ρ_k and summing over k, we then get

$$\sum_{k=1}^{n} \frac{\partial F^-}{\partial \bar{z}_k}\rho_k = \sum_{k=1}^{n} \frac{\partial F}{\partial \bar{z}_k}\rho_k - \sum_{k=1}^{n} \rho_k\rho_{\bar{k}} \sum_{m=1}^{n} \frac{\partial F}{\partial \bar{z}_m}\rho_m = 0,$$

since

$$\sum_{k=1}^{n} \rho_k\rho_{\bar{k}} = 1.$$

Next, consider the form

$$\mu_{F^-} = \sum_{k=1}^{n} (-1)^{k-1} \frac{\partial F^-}{\partial \bar{z}_k} dz[k] \wedge d\bar{z}.$$

Recall that

$$dz[k] \wedge d\bar{z}\,|_{\partial D} = (2i)^n(-1)^{n+k-1}\rho_k d\sigma,$$

where $d\sigma$ is the surface element of ∂D. Hence

$$\mu_{F^-}|_{\partial D} = (-2i)^n \sum_{k=1}^{n} \frac{\partial F^-}{\partial \bar{z}_k}\rho_k d\sigma = 0.$$

By Stokes' formula, we deduce that

$$0 = \int_{\partial D} \bar{F}^- \mu_{F^-} = (-2i)^n \int_{CD} \sum_{k=1}^{n} \left| \frac{\partial F^-}{\partial \bar{z}_k} \right|^2 dv; \qquad (17.16)$$

this is permissible, since $F^- = O(|z|^{1-2n})$, and $\partial F^-/\partial \bar{z}_k = O(|z|^{2-2n})$ as $|z| \to \infty$. (17.16) means that F^- is holomorphic outside D. Since ∂D is connected, Hartogs' theorem now shows that F^- extends holomorphically into D. However, $F^- \to 0$ as $|z| \to \infty$; hence $F^- \equiv 0$. By Theorem 2.6, F^+ extends continuously to $\bar{D}$ and $F^+|_{\partial D} = F$.

PROOF OF THEOREM 17.10. Let z^0 and $z^\pm$ be as in Lemma 17.11. Apply Lemma 17.5. In our case $N_F^+ \equiv 0$, and so N_F^- extends continuously to CD and $N_F^-|_{\partial D} = 0$. The proof proceeds from this point exactly in the same way as that of Theorem 17.9.

Note that neither of the Theorems 17.9 or 17.10 contains the other. They can be regarded as results concerning the possibility of extending F from ∂D holomorphically into D.

§18. The traces of holomorphic functions on the Shilov boundary of a circular domain

Let D be a bounded strongly starlike circular domain in $\mathbf{C}^n$, i.e. a domain invariant under all rotations $z \to e^{i\theta}z$, $0 \leqslant \theta \leqslant 2\pi$, such that $\lambda D = \{\lambda z: z \in D\} \Subset D$ for any λ, $0 < \lambda < 1$. Let S denote the Shilov boundary of D, i.e. the smallest closed set lying in $\bar{D}$ and such that

$$\sup\{|f(z)| : z \in D\} = \sup\{|f(z)| : z \in S\} \quad \text{for all } f \in AC(D).$$

If D is strictly pseudoconvex, then $S = \partial D$; in general S is a subset of ∂D.

Before proceeding to the description of the traces on S of functions from $AC(D)$, we give a characterization of polynomials orthogonal to $AC(D)$. Let μ be a nonnegative measure on S, invariant under the rotations $z \to e^{i\theta}z$, $0 \leqslant \theta \leqslant 2\pi$, and massive on S, i.e. $\overline{S \setminus E} = S$ for every $E \subset S$ with $\mu(E) = 0$. Denote by $[AC(D)]_\mu^\perp$ the space of functions $\varphi \in C(S)$ such that $\int_S \varphi f d\mu = 0$ for all $f \in AC(D)$.

If D is the unit ball and $\mu = \omega(\bar{z}, z)$, then $[AC(D)]_\mu^\perp = O(D)$.

Let $P^2(\mu)$ be the space of polynomials in $z_1, \dots, z_n$, with the inner product

$$(f, g) = \int_S f\bar{g}\,d\mu.$$

If the multi-indices I and J are such that $\|I\| \neq \|J\|$, then by the invariance of μ under rotations we have

$$\begin{aligned}(z^I, z^J) &= \int_S z^I \bar{z}^J d\mu = e^{i\theta(\|I\| - \|J\|)} \int_S z^I \bar{z}^J d\mu \\ &= e^{i\theta(\|I\| - \|J\|)} (z^I, z^J).\end{aligned}$$

Hence $(z^I, z^J) = 0$; thus, if f and g are any homogeneous polynomials of different degrees, then

$$(f, g) = 0. \tag{18.1}$$

If follows that we can choose a complete orthonormal system of polynomials $\{\varphi_l^k\}$ in $P^2(\mu)$, where l is the degree of the homogeneous polynomial, and k its index among those of degree l. For multi-indices I and J with $\|I\| \leqslant \|J\|$, we put

$$P_{I,J}(z) = \sum_k a_{IJk}\varphi_{\|J\|-\|I\|}(z),$$

where

$$a_{IJk} = \int_S z^I \bar{z}^J \varphi^k_{\|J\|-\|I\|}\, d\mu.$$

LEMMA 18.1. *A polynomial $P(z, \bar{z})$ lies in $[AC(D)]^{\perp}_{\mu}$ if and only if it is a linear combination of polynomials of the form*

$$z^I \bar{z}^J, \qquad \|I\| > \|J\|, \tag{18.2}$$

$$z^I \bar{z}^J - \overline{P_{IJ}(z)}, \qquad \|I\| \leqslant \|J\|. \tag{18.3}$$

PROOF. *Sufficiency.* D is strongly star-like, so functions from $AC(D)$ can be approximated uniformly on $\bar{D}$ by polynomials. Since $\{\varphi_l^k\}$ is complete, we need only prove that, for a polynomial $P(z, \bar{z})$ of the form (18.2) or (18.3) and for all k and l,

$$\int_S P(z, \bar{z})\varphi_l^k d\mu = 0. \tag{18.4}$$

If $P(z, \bar{z})$ is of the form (18.2), then by (18.1)

$$\int_S P(z, \bar{z})\varphi_l^k d\mu = (z^I\varphi_l^k, z^J) = 0,$$

since $z^I\varphi_l^k$ is a homogeneous polynomial of degree $\|I\| + l > \|J\|$. Next, if $P(z, \bar{z})$ is of the form (18.3) and $l \neq \|J\| - \|I\|$, we have (using (18.1) again)

$$\int_S P(z, \bar{z})\varphi_l^k d\mu = (z^I\varphi_l^k, z^J) - (\varphi_l^k, P_{IJ}(z)) = 0.$$

Finally, if $l = \|J\| - \|I\|$, then (18.4) holds by the choice of the a_{IJk}.

Necessity. Let $P(z, \bar{z}) = \sum_{I,J} b_{IJ} z_I \bar{z}_J$ belong to $[AC(D)]^{\perp}_{\mu}$; then

$$P(z, \bar{z}) = \sum_{\|I\|>\|J\|} b_{IJ} z^I \bar{z}^J + \sum_{\|I\|\leqslant\|J\|} b_{IJ}\left(z^I\bar{z}^J - \overline{P_{IJ}(z)}\right) + \overline{\sum_{\|I\|\leqslant\|J\|} \bar{b}_{IJ} P_{IJ}(z)},$$

i.e. $P(z, \bar{z}) = P_1(z, \bar{z}) + P(z)$, where P_1 is a linear combination of polynomials of the form (18.2) or (18.3), and $P(z)$ is a holomorphic polynomial. Since $P(z, \bar{z}) \in [AC(D)]_\mu^\perp$ by assumption, and $P_1(z, \bar{z}) \in [AC(D)]_\mu^\perp$ by what we have already proved, we see that $\overline{P(z)} \in [AC(D)]_\mu^\perp$. Hence

$$\int_S |P(z)|^2 \, d\mu = \int_S \overline{P(z)} \cdot P(z) d\mu = 0.$$

Since μ is a nonnegative and massive on S, it follows that $P(z) \equiv 0$ on S, and hence that $P(z) \equiv 0$ on $\bar{D}$ (S is the Shilov boundary). Thus $P(z, \bar{z}) = P_1(z, \bar{z})$. Q.E.D.

If D is n-circular, i.e. invariant under all n-rotations $z \to (e^{i\theta_1} z_1, \ldots, e^{i\theta_n} z_n)$, $0 \leqslant \theta_1, \ldots, \theta_n \leqslant 2\pi$, and if μ is also invariant under n-rotations, then as a complete orthonormal system we an choose $\{\alpha_I z^I\}$, consisting of monomials with $\alpha_I = (\int_S z^I \bar{z}^I \, d\mu)^{-1/2}$.

In this case Lemma 18.1 becomes

LEMMA 18.2. *The polynomial $P(z, \bar{z})$ lies in $[AC(D)]_\mu^\perp$ if and only if it is a linear combination of polynomials of the form*

$$z^I \bar{z}^J, \quad \textit{with } i_k > j_k \textit{ for some } k,$$

$$z^I \bar{z}^{I+J} \left(\alpha_{I+J}^2 - \alpha_{I+J+e_k}^2 |z_k|^2 \right); \tag{18.5}$$

where e_k is the multi-index with 1 *at the kth place and* 0 *elsewhere.*

Lemmas 18.1 and 18.2 are analogues of Lemma 5.7. When D is the unit ball, $w = \bar{z}$ and $\mu = \omega(z, \bar{z})$, Lemma 18.2 and Lemma 5.7 are identical.

For the proof of the main result, we need the Poisson kernel. First we construct the Szegö kernel $h(\bar{\zeta}, z)$ of the circular domain D by the formula

$$h(\bar{\zeta}, z) = \sum_{k,l} \varphi_l^k(z) \overline{\varphi_l^k(\zeta)}; \qquad z \in D, \zeta \in \bar{D}.$$

It can be shown that:

1) $h(\bar{\zeta}, z)$ is holomorphic in z on D and continuous in ζ on $\bar{D}$;
2) $h(\bar{\zeta}, z) = \overline{h(\bar{z}, \zeta)}$ on $D \times D$; and
3) for all $f \in AC(D)$,

$$f^{(z)} = \int_S f(\zeta) h(\bar{\zeta}, z) d\mu_\zeta. \tag{18.6}$$

The Poisson kernel can be constructed from the Szegö kernel:

$$P(\zeta, z) = \frac{|h(\bar{\zeta}, z)|^2}{h(\bar{z}, z)}.$$

For $z \in D$ we have $h(\bar z, z) > 0$, since (18.6) and 2) above imply

$$h(\bar z, z) = \int_S h(\bar z, \zeta)h(\bar\zeta, z)\,d\mu_\zeta = \int_S |h(\bar\zeta, z)|^2\,d\mu_\zeta > 0.$$

Further, if $f \in AC(D)$, then

$$f(z) = \frac{1}{h(\bar z, z)} f(z)h(\bar z, z) = \frac{1}{h(\bar z, z)} \int_S f(\zeta)h(\bar z, \zeta)h(\bar\zeta, z)\,d\mu_\zeta$$

$$= \int_S f(\zeta)P(\zeta, z)\,d\mu_\zeta. \tag{18.7}$$

THEOREM 18.3. *Let D be a bounded strongly star-like circular domain in $\mathbf{C}^n$ such that every point of its Shilov boundary S is a peak-point for the algebra $AC(D)$, and let μ be a nonnegative massive measure on S invariant under the rotations $z \to e^{i\theta}z$, $0 \leqslant \theta \leqslant 2\pi$. Then a function $\varphi \in C(S)$ is of the form $f|_S$ for some $f \in AC(D)$ if and only if*

$$\int_S \varphi P\,d\mu = 0 \tag{18.8}$$

for every polynomial P of the form (18.2) *or* (18.3).

If D is n-circular and μ is invariant under all the n-rotations $z \to (e^{i\theta_1}z_1, \ldots, e^{i\theta_n}z_n)$, $0 \leqslant \theta_1, \ldots, \theta_n \leqslant 2\pi$, then we may replace (18.2) and (18.3) above by (18.5).

PROOF. The necessity follows from Lemma 18.1.

Sufficiency. Since each point of S is a peak-point for $AC(D)$, we have for each $z^0 \in S$ and any neighborhood W of z^0

$$\lim_{\substack{z \to z^0 \\ z \in D}} \int_{S \setminus w} P(\zeta, z)\,d\mu_\zeta = 0. \tag{18.9}$$

The argument for this can be found in Koráni [1]. Further, since $P(\zeta, z) \geqslant 0$ and

$$\int_S P(\zeta, z)\,d\mu_\zeta = 1, \qquad z \in D,$$

it follows from (18.9) (see, for example, Hoffman [1], p. 18) that if $\varphi \in C(S)$, then the function

$$f(z) = \begin{cases} \varphi(z), & z \in S, \\ \displaystyle\int_S \varphi(\zeta)P(\zeta, z)\,d\mu_\zeta, & z \in D, \end{cases}$$

is continuous on $D \cup S$. Since S is the Shilov boundary of D, $\{f(rz)\}$ therefore converges uniformly on $\bar{D}$ as $r \to 1$; hence f extends continuously to $\bar{D}$. It remains to show that, if φ satisfies (18.8), then $f \in A(D)$. From (18.7), we see that, for any $g \in AC(D)$ and any differential operator T with constant coefficients,

$$\int_S g(\zeta) T_z \bar{\partial}_z (P(\zeta, z)) d\mu_\zeta \equiv 0.$$

It follows in particular that

$$T_z \bar{\partial}_z (P(\zeta, z)) \big|_{z=0} \in [AC(D)]^\perp_\mu .$$

From the form of the Poisson and Szegö kernels, we see that the function on the left above is a polynomial in ζ and $\bar{\zeta}$. By Lemma 18.1, it is a linear combination of polynomials of the form (18.2) and (18.3). This and the hypothesis of the theorem imply that

$$\int_S \varphi(\zeta) [T_z \bar{\partial}_z (P(\zeta, z))] \big|_{z=0} d\mu_\zeta = 0,$$

i.e. all the coefficients of the form

$$\bar{\partial} f(z) = \int_S \varphi(\zeta) \bar{\partial}_z P(\zeta, z) d\mu_\zeta \tag{18.10}$$

and all their derivatives vanish at $z = 0$. Since the Poisson kernel is real-analytic in z, the form (18.10) has real-analytic coefficients; hence $\bar{\partial} f \equiv 0$ on D, i.e. $f \in A(D)$.

§19. Computation of an integral of Martinelli-Bochner type for the case of the ball

We consider on the unit sphere $\partial B(0, 1)$ the space $L^2(\partial B(0, 1))$ with the usual inner product

$$(f, g) = \int_{\partial B(0,1)} f \cdot \bar{g} \, d\sigma,$$

where $d\sigma$ is the surface element of the sphere. We shall identify this space with the space $G^2(B(0, 1))$ of harmonic functions f in $B(0, 1)$ represented by a Poisson integral from $L^2(\partial B(0, 1))$.

Let $\mathcal{P}_{s,t}$ be the finite-dimensional vector space consisting of homogeneous harmonic polynomials of the form

$$P_{s,t} = \sum_{\|I\|=s} \sum_{\|J\|=t} a_{IJ} z^I \bar{z}^J.$$

The union of all the $\mathcal{P}_{s,t}$ is dense in $L^2(\partial B(0,1))$. As we know, (17.1) is valid for $f \in G^2(B(0,1)) \cap A(B(0,1))$:

$$(Mf)(z) = \int_{\partial B(0,1)} f(\zeta) U_{0,0}(\zeta, z) = f(z), \qquad z \in B(0,1).$$

LEMMA 19.1 *Let $P_{s,t} \in \mathcal{P}_{s,t}$. Then the Martinelli-Bochner type integral of $P_{s,t}$ is of the form*

$$MP_{s,t} = \frac{n+s-1}{n+s+t-1} P_{s,t}.$$

This follows from the fact that the harmonic extension of $\zeta_k P_{s,t}$ from $\partial B(0,1)$ into $B(0,1)$ is given by

$$\zeta_k P_{s,t} + \frac{1-|\zeta|^2}{n+s+t-1} \frac{\partial}{\partial \bar\zeta_k} P_{s,t},$$

and from the easily verified formula

$$U_{0,0}(\zeta, z)\,|_{\partial B(0,1)} = \frac{1-\langle \zeta, \bar z\rangle}{1-|z|^2} P(\zeta, z)\,d\sigma;$$

here $P(\zeta, z)$ is the Poisson kernel for the ball:

$$P(\zeta, z) = \frac{(n-1)!}{2\pi^n} \frac{1-|z|^2}{|\zeta - z|^{2n}}.$$

COROLLARY 19.2. *If*

$$Q_{s,t} = \sum_{\|I\|=s} \sum_{\|J\|=t} a_{IJ} z^I \bar z^J$$

then

$$MQ_{s,t} = \sum_{p \geqslant 0} R_{s-p,t-p},$$

where the $R_{s-p,t-p}$ are the harmonic polynomials defined by

$$R_{s-p,t-p} = \frac{s+p+n-1}{p!\,(s+t+n-p-1)!} \times \sum_{j \geqslant 0} (-1)^j \frac{(s+t-j-2p+n-2)!}{j} |z|^{2j} \Delta^{j+p} Q_{s,t},$$

Δ being the Laplace operator:

$$\Delta R = \sum_{j=1}^{n} \frac{\partial^2 R}{\partial z_j \partial \bar z_j}.$$

This corollary follows from Lemma 19.1 and the Gauss representation of an arbitrary polynomial on the sphere as a sum of homogeneous harmonic polynomials (see, for example, Sobolev [1], Chapter 13).

We introduce the differential operators b and $\bar{b}$ by setting

$$b = \sum_{k=1}^{n} z_k \frac{\partial}{\partial z_k}, \qquad \bar{b} = \sum_{k=1}^{n} \bar{z}_k \frac{\partial}{\partial \bar{z}_k}.$$

THEOREM 19.3. *Suppose $f \in G^2(B(0,1))$. Then the Martinelli-Bochner type integral of f has the form*

$$Mf = f - \bar{b}\psi = f - \sum_{k=1}^{n} \bar{z}_k \psi_k, \tag{19.1}$$

where

$$\psi = |z|^{1-n} \int_0^{|z|} |\zeta|^{n-2} f(\zeta)\, d|\zeta|,$$

$$\psi_k = |z|^{-n} \int_0^{|z|} |\zeta|^{n-1} \frac{\partial}{\partial \bar{\zeta}_k} f(\zeta)\, d|\zeta|, \qquad k = 1, \ldots, n.$$

PROOF. Since $f \in G^2(B(0,1))$, in the interior of $B(0,1)$ an expansion

$$f(\zeta) = \sum_{s,t \geqslant 0} P_{s,t}(\zeta), \tag{19.2}$$

converging uniformly on compact subsets of $B(0,1)$, in particular on each sphere $\partial B(0,r)$, $0 < r < 1$. From the definition of the class $G^2(B(0,1))$, it is not hard to show that (19.2) is also valid on $\partial B(0,1)$, but in the sense of convergence in $L^2(\partial B(0,1))$. By Lemma 19.1, we deduce that, in the interior of $B(0,1)$,

$$Mf = \sum_{s,t \geqslant 0} \frac{n+s-1}{n+s+t-1} P_{s,t} = f - \sum_{s,t \geqslant 0} \frac{t}{n+s+t-1} P_{s,t}.$$

To prove the first part of (19.1), we observe that

$$\psi(z) = |z|^{1-n} \sum_{s,t \geqslant 0} \int_0^{|z|} |\zeta|^{n-2+t+s} P_{s,t}\left(\frac{\zeta}{|\zeta|}\right) d|\zeta|$$

$$= \sum_{s,t \geqslant 0} \frac{|z|^{t+s}}{n+s+t-1} P_{s,t}\left(\frac{z}{|z|}\right) = \sum_{s,t \geqslant 0} \frac{1}{n+s+t-1} P_{s,t},$$

$$\bar{B}\psi = \sum_{s,t \geqslant 0} \frac{1}{n+s+t-1} \sum_{k=1}^{n} \bar{z}_k \frac{\partial P_{s,t}}{\partial \bar{z}_k} = \sum_{s,t \geqslant 0} \frac{t}{n+s+t-1} P_{s,t}.$$

The proof of the second part of (19.1) is similar.

The above theorem enables us to compute Mf for any $f \in G^2(B(0,1))$. The functions ψ and ψ_k are harmonic in $B(0,1)$. To compute them, it is convenient to change over to polar coordinates. If f is polyharmonic or real-analytic in $B(0,1)$, then it may be convenient to compute Mf by using the Almani representation of f (see Sobolev [1], Chapter 14):

$$f = \sum_{k\geqslant 0} f_k|z|^{2k},$$

with the f_k harmonic in $B(0,1)$.

We present some examples to illustrate the applications of Theorem 19.3.

EXAMPLE 1. Let $f \in G^2(B(0,1))$, and

$$Tf = \frac{(n-2)!}{(2\pi)^n}\int_{\partial B(0,1)} f(\zeta)\frac{d\sigma_\zeta}{|\zeta - z|^{2n-2}}, \qquad z \in \overline{B(0,1)},$$

the potential of a simple layer. Then

$$Tf = |z|^{1-n}\int_0^{|z|} |\zeta|^{n-2} f(\zeta)\,d|\zeta|.$$

In particular, if $Tf = F$, then $f = (n-1)F + bF + \bar{b}F$.

Consider the integral equation

$$f + \lambda Tf = \varphi, \qquad \varphi \in G^2(B(0,1)) \tag{19.3}$$

on $\partial B(0,1)$. Then its solution is of the form

$$f = \varphi - \lambda|z|^{1-n-\lambda}\int_0^{|z|} |\zeta|^{\lambda+n-2}\varphi(\zeta)\,d|\zeta|, \qquad |\lambda| < n-1. \tag{19.4}$$

EXAMPLE 2. Let $f \in L^2(\partial B(0,1))$, and

$$Sf = 2\int_{\partial B(0,1)} f(\zeta)U_{0,0}(\zeta, z), \qquad z \in \partial B(0,1),$$

be the singular integral of the Martinelli-Bochner type. Consider the singular equation

$$f + \lambda Sf = \varphi, \qquad \varphi \in L^2(\partial B(0,1)). \tag{19.5}$$

If φ has an expansion $\varphi = \Sigma_{s=0} P_{ps,qs}$ in $B(0,1)$, where p and q are nonnegative integers, then

$$f = \frac{p+q}{(1-\lambda)q + (1+\lambda)p}\varphi - \frac{2\lambda q(p+q)(n-1)}{[(1-\lambda)q+(1+\lambda)p]^2}|z|^{-\beta}\int_0^{|z|}|\zeta|^{\beta-1}\varphi(\zeta)\,d|\zeta|, \tag{19.6}$$

$$|\lambda| < 1, \qquad \beta = \frac{(n-1)(1+\lambda)(p+q)}{(1-\lambda)q + (1+\lambda)p}$$

If we know only the values of φ on $\partial B(0,1)$, we must replace φ in (19.4) and (19.6) by its Poisson integral. Then we obtain integral representations for the solutions of (19.3) and (19.5), i.e. resolvents of the operators T and S.

To compute integrals similar to Mf in more general domains than the ball, we may use the ideas of §5. We will have to require that (5.8) holds, or that the polynomials $P(z, w)$ are dense in $L^2(\partial D)$.

We indicate other applications of Lemma 19.1. The following theorem is an immediate consequence of the lemma.

THEOREM 19.4. *For $n > 1$ the operator M maps $L^2(\partial B(0,1))$ into itself, is bounded, and has norm* 1. *All positive rational numbers in* $[0,1]$ *are eigenvalues of infinite multiplicity for M. The spectrum of M is* $[0,1]$.

Denote by Pr_H the orthogonal projection of $G^2(B(0,1))$ into the Hardy space $H^2(B(0,1))$ consisting of elements of $A(B(0,1))$ satisfying the condition that $\int_{\partial B(0,r)} |f|^2\,d\sigma$ is uniformly bounded for $0 < r < 1$. This operator is called Szegö's operator, and has the integral representation

$$(\mathrm{Pr}_H f)(z) = \int_{\partial B(0,1)} f(\zeta) \frac{\sum_{k=1}^n (-1)^{k-1} \bar{\zeta}_k d\bar{\zeta}\,[k] \wedge d\zeta}{(1 - \langle \bar{\zeta}, z \rangle)^n}.$$

For $n = 1$, integrals of Martinelli-Bochner type are integrals of Cauchy type, and $M = \mathrm{Pr}_H$. For $n > 1$, the kernel $U_{0,0}$ is not holomorphic, and M and Pr_H are not the same, as the following example shows.

EXAMPLE 3. Let $f = \bar{z}_k$. Then f^+ is not holomorphic in $B(0,1)$. Indeed,

$$\begin{aligned} &\frac{\partial}{\partial \bar{z}_k} f^+(z) \\ &= \frac{(n-1)!}{(2\pi i)^n} \int_{\partial B(0,1)} \bar{\zeta}_k \Bigg[n \sum_{m=1}^n (-1)^{m-1} \frac{(\bar{\zeta}_m - \bar{z}_m)(\zeta_k - z_k)}{|\zeta - z|^{2n+2}} d\bar{\zeta}\,[m] \wedge d\zeta \\ &\qquad - (-1)^{k-1} \frac{1}{|\zeta - z|^{2n}} d\bar{\zeta}\,[k] \wedge d\zeta \Bigg]. \end{aligned}$$

Since

$$U_{0,0}(\zeta, z)\,|_{\partial B(0,1)} = \frac{(n-1)!}{2\pi^n} \sum_{m=1}^n \frac{(\bar{\zeta}_m - \bar{z}_m)\zeta_m}{|\zeta - z|^{2n}} d\sigma,$$

we have

$$\frac{\partial}{\partial z_k} f^+ \,|_{z=0} = \frac{(n-1)!}{2\pi^n} (n-1) \int_{\partial B(0,1)} |\zeta_k|^2\, d\sigma,$$

and the integral vanishes only if $n = 1$.

However, the following theorem is true:

THEOREM 19.5. *For every* $f \in L^2(\partial B(0,1))$,

$$\lim_{k\to\infty} M^k f = \mathrm{Pr}_H f, \tag{19.7}$$

where M^k *denotes the kth iterate of* M.

PROOF. For each homogeneous harmonic polynomial $P_{s,t}$, (19.7) follows immediately from Lemma 19.1, since

$$\mathrm{Pr}_H P_{s,t} = \begin{cases} P_{s,t}, & \text{if } t = 0, \\ 0, & \text{if } t > 0. \end{cases}$$

But such polynomials are dense in $G^2(B(0,1))$, and the family of iterates of M is uniformly bounded in norm; hence (see Dunford and Schwartz [1], Chapter II, §3, Theorem 6) (19.7) holds on all of $L^2(B(0,1))$.

§20. Differential boundary conditions for the holomorphy of functions

Let $D = \{z \in \mathbf{C}^n\colon \rho(z) < 0\}$ be a bounded domain with $\partial D \in C^1$. We consider the following problem: suppose we are given a function $f \in C^1(\bar D)$ harmonic in D, and a continuous vector field $w = (w_1(z),\dots,w_n(z))$ on ∂D, with

$$\langle w, \overline{\mathrm{grad}\,\rho}\rangle = \sum_{j=1}^n W_j \frac{\partial\rho}{\partial z_j} \neq 0, \qquad z \in \partial D, \tag{20.1}$$

i.e. the vector $\bar w$ never lies in the complex tangent plane $T^c_{\partial D}(z)$ at z to ∂D for any $z \in \partial D$. Then does the condition

$$\sum_{k=1}^n \frac{\partial f}{\partial \bar z_k} w_k = 0, \qquad z \in \partial D, \tag{20.2}$$

imply that f is holomorphic in D? This problem is an analogue of the oblique derivative problem for real-valued harmonic functions. If we do not require (20.1), then there does exist an $f \notin A(D)$ satisfying (20.2), e.g. $D = B(0,1) \subset \mathbf{C}^2$, $w = (0,1)$ and $f = \bar z_1$.

When $w = \mathrm{grad}\,\rho$, (20.2) becomes the Neumann normal $\bar\partial$-condition for functions harmonic in D, and the answer to our question is affirmative:

THEOREM 20.1. *Let* D *be a bounded domain in* $\mathbf{C}^n$, *and let* $f \in C(\bar D)$ *be harmonic in* D. *Then* $f \in A(D)$ *if and only if the* $\partial f/\partial \bar z_k$, $k = 1,\dots,n$, *can be extended continuously to* $\bar D$ *and satisfy the Neumann normal* $\bar\partial$*-condition*

$$\sum_{k=1}^n \frac{\partial f}{\partial \bar z_k}\cdot\frac{\partial\rho}{\partial z_k} = 0, \qquad z \in \partial D. \tag{20.3}$$

PROOF. If $f \in A(D)$, then all $\partial f/\partial \bar{z}_k \equiv 0$, and (20.3) is trivial. Conversely, if $\partial f/\partial \bar{z}_k \in C(\bar{D})$, $k = 1,\ldots,n$, and (20.3), holds, then

$$\mu_f|_{\partial D} = (-2i)^n \sum_{k=1}^{n} \frac{\partial f}{\partial \bar{z}_k} \cdot \frac{\partial \rho}{\partial z_k} d\sigma = 0$$

where $d\sigma$ is the surface element of ∂D. As in the proof of Theorem 17.3, we now conclude easily that $f \in A(D)$.

Let us reformulate our problem differently. Decompose w into its normal and tangential components to ∂D at a point $z \in \partial D$:

$$w(z) = \alpha(z)\operatorname{grad} \rho + \tau(z);$$

here α never vanishes on ∂D, and $\tau(z)$ is orthogonal to $\operatorname{grad} \rho$, i.e. $\langle \overline{\operatorname{grad} \rho}, \tau \rangle \equiv 0$ on ∂D. Then (20.2) becomes

$$\sum_{k=1}^{n} \frac{\partial f}{\partial \bar{z}_k} \frac{\partial \rho}{\partial z_k} = -\sum_{k=1}^{n} \frac{\partial f}{\partial \bar{z}_k} \cdot \frac{\tau_k(z)}{\alpha(z)}. \tag{20.4}$$

Consider again the form

$$\mu_f = \sum_{k=1}^{n} (-1)^{k-1} \frac{\partial f}{\partial \bar{z}_k} dz[k] \wedge d\bar{z}.$$

Using (20.4), we can write it on ∂D as

$$\mu_f|_{\partial D} = \sum_{k>m} a_{k,m}(z) df \wedge d\bar{z}[k,m] \wedge dz|_{\partial D}, \tag{20.5}$$

where the $a_{k,m} \in C(\partial D)$. Then the new formulation of our problem is: suppose $f \in C^1(\bar{D})$ is harmonic in D, and (20.5) holds; does it follow that f is holomorphic in D? (20.5) reduces to the Neumann normal $\bar{\partial}$-condition when all the $a_{k,m}$ are zero. Condition (20.5) can also be written in integral form:

$$f(z) = \int_{\partial D} f(\zeta)\left[U_{0,0}(\zeta, z) + (-1)^n \sum_{k>m} d(g a_{k,m}) \wedge d\bar{\zeta}\,[k,m] \wedge d\zeta\right] = 0$$

with $g(\zeta, z) = -(n-2)!(2\pi i)^{-n}|\zeta - z|^{2-2n}$. It is then clear that the problem of the holomorphy of the functions defined by a Martinelli-Bochner integral (see Theorems 17.1 and 17.2) is a special case of the present problem (the case $a_{k,m} = 0$).

We now present the solution to our problem for certain classes of domains D and functions $a_{k,m}$.

THEOREM 20.2. *If $n = 2$, and ∂D is connected and of class C^1, and a harmonic function $f \in C^1(\bar{D})$ satisfies the condition*

$$\mu_f|_{\partial D} = a\, df \wedge d\zeta|_{\partial D}, \tag{20.6}$$

where a is a constant, then $f \in A(D)$.

Proof. If we show that, for each $P_{s,t} \in \mathcal{P}_{s,t}$ (see §19 for the notation), $s, t = 0, 1, 2, \ldots$,

$$\int_{\partial D} f(\zeta) \sum_{k=1}^{2} (-1)^{k-1} \frac{\partial P_{s,t}}{\partial \zeta_k} d\bar\zeta\,[k] \wedge d\zeta = 0, \tag{20.7}$$

holds, then our assertion will follow from Corollary 17.4. By (20.6) and Stokes' theorem,

$$\int_{\partial D} f(\zeta) \sum_{k=1}^{2} (-1)^{k-1} \frac{\partial P_{s,t}}{\partial \zeta_k} d\bar\zeta\,[k] \wedge d\zeta = \int_{\partial D} P_{s,t}\mu_f$$

$$= \int_{\partial D} aP_{s,t} df \wedge d\zeta = -\int_{\partial D} af\, dP_{s,t} \wedge d\zeta.$$

If $t = 0$, i.e. $P_{s,t}$ is a holomorphic polynomial, (20.7) holds. Let $t > 0$. We transform $dP_{s,t} \wedge d\zeta$ into the form

$$\frac{\partial \tilde P}{\partial \zeta_1} d\bar\zeta_2 \wedge d\zeta - \frac{\partial \tilde P}{\partial \zeta_2} d\bar\zeta_1 \wedge d\zeta.$$

This can be achieved by setting

$$\tilde P = -\int \frac{\partial P_{s,t}}{\partial \bar\zeta_1} d\zeta_2$$

(formal integration with respect to ζ_2). Then

$$\frac{\partial \tilde P}{\partial \zeta_1} = -\int \frac{\partial^2 P_{s,t}}{\partial \zeta_1 \partial \bar\zeta_1} = \int \frac{\partial^2 P_{s,t}}{\partial \zeta_2 \partial \bar\zeta_2} d\zeta_2 = \frac{\partial P_{s,t}}{\partial \bar\zeta_2},$$

so that $\Delta \tilde P = 0$. Also, $\tilde P \in \mathcal{P}_{s+1,t-1}$. Thus

$$\int_{\partial D} f(\zeta) \sum_{k=1}^{2} (-1)^{k-1} \frac{\partial P_{s,t}}{\partial \zeta_k} d\bar\zeta\,[k] \wedge d\zeta$$

$$= a\int_{\partial D} f(\zeta) \sum_{k=1}^{2} (-1)^{k-1} \frac{\partial \tilde P}{\partial \zeta_k} d\bar\zeta\,[k] \wedge d\zeta. \tag{20.8}$$

Now we apply induction on t, using the fact that (20.7) holds for $t = 0$, and conclude from (20.8) that (20.7) is valid for all t.

Theorem 20.3. *Let $D = B(0, 1)$ and suppose that the $a_{k,m}$ lie in $AC(D)$. Then, if $f \in C^1(\bar D)$ is harmonic in D and satisfies* (20.5), *it follows that $f \in A(D)$.*

Proof. Consider first the case

$$\mu_f\,|_{\partial B(0,1)} = Q_{l,0}(\zeta)\partial f \wedge d\bar\zeta\,[k,m] \wedge d\zeta\,|_{\partial B(0,1)}, \tag{20.9}$$

where $Q_{l,0} \in \mathcal{P}_{l,0}$.

Writing out (20.9) for the case of the ball, we have

$$\sum_{j=1}^{n} \frac{\partial f}{\partial \bar{\zeta}_j} \bar{\zeta}_j = \pm Q_{l,0}(\zeta) \left[\frac{\partial f}{\partial \bar{\zeta}_k} \zeta_m - \frac{\partial f}{\partial \bar{\zeta}_m} \zeta_k \right], \qquad \zeta \in \partial B(0,1). \quad (20.10)$$

We shall look for the solution of (20.10) in the form of a series

$$f = \sum_{s,t \geq 0} P_{s,t},$$

where $P_{s,t} \in \mathcal{P}_{s,t}$. Then (20.10) yields

$$\sum_{s,t \geq 0} t\mathcal{P}_{s,t} = \pm Q_{l,0}(\zeta) \sum_{s,t \geq 0} \left(\zeta_m \frac{\partial P_{s,t}}{\partial \bar{\zeta}_k} - \zeta_k \frac{\partial P_{s,t}}{\partial \bar{\zeta}_m} \right). \quad (20.11)$$

We now extend the function on the right above harmonically into $B(0,1)$ using Gauss' formula (see, for example, Sobolev [1], Chapter 11): if P_k is any homogeneous polynomial of degree k, then its harmonic extension P_Γ from $\partial B(0,1)$ into $B(0,1)$ is given by

$$P_\Gamma = \sum_{s \geq 0} Z_{k-2s},$$

where

$$Z_{k-2s} = \frac{k-2s+n-1}{s!\,(k+n-s-1)!} \sum_{j \geq 0} (-1)^j \frac{(k-j-2s+n-2)!}{j!} |z|^{2j} \Delta^{j+s} P_k. \quad (20.12)$$

In our case

$$P_k = Q_{l,0} \left(\zeta_m \frac{\partial P_{s,t}}{\partial \bar{\zeta}_k} - \zeta_k \frac{\partial P_{s,t}}{\partial \bar{\zeta}_m} \right),$$

and $k = l + s + t$. Hence (20.11) leads to the system of equations

$$\begin{aligned} tP_{s,t} = {} & \sum_{j \geq 0} a_j |\zeta|^{2j} \Delta^j \left[Q_{l,0} \left(\frac{\partial P_{s-l-1,t+1}}{\partial \bar{\zeta}_k} \zeta_m - \frac{\partial P_{s-l-1,t+1}}{\partial \bar{\zeta}_m} \zeta_k \right) \right] \\ & + \sum_{j \geq 0} b_j |\zeta|^{2j} \Delta^{j+1} \left[Q_{l,0} \left(\frac{\partial P_{s-l,t+2}}{\partial \bar{\zeta}_k} \zeta_m - \frac{\partial P_{s-l,t+2}}{\partial \bar{\zeta}_m} \zeta_k \right) \right] + \cdots \\ & + \sum_{j \geq 0} c_j |\zeta|^{2j} \Delta^{j+l} \left[Q_{l,0} \left(\frac{\partial P_{s-1,t+l+1}}{\partial \bar{\zeta}_k} \zeta_m - \frac{\partial P_{s-1,t+l+1}}{\partial \bar{\zeta}_m} \zeta_k \right) \right], \end{aligned} \quad (20.13)$$

where the $a_j, b_j, \ldots, c_j$ are constants. For $s = 0$, we have $tP_{0,t} = 0$ for any t by (20.13); hence $P_{0,t} = 0$ for $t \neq 0$. Next we have $tP_{1,t} = 0, \ldots, tP_{s,t} = 0$, so that the series for f consists only of holomorphic summands, i.e. f is holomorphic in $B(0,1)$.

If the $a_{k,m}$ are arbitrary holomorphic functions, we can decompose them into sums of homogeneous polynomials and apply the earlier considerations. Then we get a system of equations of the type of (20.13), with each equation involving only finitely many summands.

In the case of $B(0, r)$, the Neumann normal $\bar{\partial}$-condition has the form

$$\sum_{k=1}^{n} \bar{z}_k \frac{\partial f}{\partial \bar{z}_k} = 0 \quad \text{on } \partial B(0, r).$$

Suppose now that this condition is satisfied everywhere in $B(0, r)$. Then it turns out that the requirement that f be harmonic can be dropped, and f will still be holomorphic in $B(0, r)$ as before:

THEOREM 20.4. *Let $f \in C^1(B(0, r)) \cap C^\infty(0)$, and suppose that*

$$\sum_{k=1}^{n} \bar{z}_k \frac{\partial f}{\partial \bar{z}_k} = 0, \qquad z \in B(0, r). \tag{20.14}$$

Then f is holomorphic in $B(0, r)$. If $f \in C^2(B(, r)) \cap C^\infty(0)$ and

$$\sum_{k,m=1}^{n} \bar{z}_k z_m \frac{\partial^2 f}{\partial \bar{z}_k \partial z_m} = 0, \qquad z \in B(0, r), \tag{20.15}$$

then f is pluriharmonic in $B(0, r)$.

PROOF. Observe that (20.14) implies that f is holomorphic on each complex line l through 0, and (20.15) implies that f is harmonic on each such line. Thus we need only appeal to a result of Forelli [1], which asserts that an f which is harmonic on each complex line l through 0 and C^∞ at 0 is pluriharmonic in $B(0, r)$.

Rudin [1] has shown that, if a function f in $B(0, r)$ satisfies the Laplace equation $\Delta f = 0$ and the Laplace-Beltrami equation $\Delta' f = 0$, where

$$\Delta' = \Delta - \sum_{k,m=1}^{n} \bar{z}_k z_m \frac{\partial^2}{\partial \bar{z}_k \partial z_m},$$

then f is pluriharmonic in $B(0, r)$. Theorem 20.4 shows that, if $f \in C^2(B) \cap C^\infty(0)$, then the single equation (20.15) is sufficient. The single condition $\Delta' f = 0$ is not sufficient. For example, let

$$f = \left(|z_1|^2 - |z_2|^2\right) \sum_{k \geq 0} \frac{k!\,(n+1)!}{(n+k+1)!} |z|^{2k}. \tag{20.16}$$

Then f is real-analytic in $B(0, 1)$, since the radius of convergence of (20.16) is 1. It is easy to verify that $\Delta' f = 0$, but f is not pluriharmonic, since

$$\sum_{k,m} \bar{z}_k z_m \frac{\partial^2 f}{\partial \bar{z}_k \partial z_m} = \left(|z_1|^2 - |z_2|^2\right) \sum_{j \geq 0} \frac{(j+1)!\,(j+1)(n+1)!}{(n+j+1)!} |z|^{2j}.$$

If $f \in C^k(B(0, r))$, Theorem 20.4 is not true; e.g. let

$$f(z) = \frac{z^I \sum_{j=1}^n |z_j|^4}{|z|^4}.$$

Since $\sum_1^n |z_j|^4 \leqslant |z|^4$, we see that, by choosing a multi-index I with $\|I\|$ sufficiently large, we can have $f \in C^k(B(0, r))$ for any prescribed k. Clearly f is not holomorphic in $B(0, r)$, although $\sum_1^n \bar{z}_k \partial f / \partial \bar{z}_k = 0$.

Another theorem related to the above theorems by the method of proof is the following:

THEOREM 20.5. *Let P_k be an arbitrary homogeneous polynomial of degree $k \geqslant 2$. Then the restriction of P_k to $\partial B(0, r)$ can be extended holomorphically into $B(0, r)$ if and only if $\mathfrak{L}_k P_k = 0$ on $\partial B(0, r)$, where*

$$\mathfrak{L}_k = \sum_{j=1}^n \bar{z}_j \frac{\partial}{\partial \bar{z}_j} - \sum_{l \geqslant 0} \frac{(n + k - 2l - 3)!\,(2l - 1)!!2^l}{(n + k - 2)!\,(l + 1)!} \Delta^{l+1}.$$

PROOF. Let $P_k = P_{m,t}$, i.e.

$$P_k = \sum_{\|I\| = m} \sum_{\|J\| = t} a_{IJ} z^I \bar{z}^J.$$

We extend P_k from $\partial B(0, r)$ into $B(0, r)$ by Gauss' formula; the extension, call it P_Γ, can be written as $P_\Gamma = \sum_{s \geqslant 0} Z_{k-2s}$, where Z_{k-2s} is defined by (20.12). P_Γ is holomorphic if and only if

$$\sum_{j=1}^n \bar{z}_j \frac{\partial P_\Gamma}{\partial \bar{z}_j} = 0$$

in $B(0, r)$ (see Theorem 20.4). But

$$\sum_{j=1}^n \bar{z}_j \frac{\partial Z_{k-2s}}{\partial \bar{z}_j} = (t - s) Z_{k-2s}.$$

Then

$$\sum_{j=1}^n \bar{z}_j \frac{\partial P_\Gamma}{\partial \bar{z}_j} = \sum_{j,s} (t - s) \frac{(-1)^j (k - 2s + n - 1)(k - j - 2s + n - 2)!}{s!\,j!\,(k + n - s - 1)!} \Delta^{j+s} P_k$$

$$= \sum_{l \geqslant 0} \Delta^l P_k \sum_{j+s=l} \frac{(-1)^j (k - 2s + n - 1)(k - l - s + n - 2)!(t - s)}{s!(l - s)!(k + n - s - 1)!}$$

$$= \sum_{l \geqslant 0} (-1)^l \Delta^l P_k \sum_{s=0}^{l} \frac{(-1)^s (k - 2s + n - 1)(k - l - s + n - 2)!(t - s)}{s!(l - s)!(k + n - s - 1)!}.$$

The following lemma now concludes the proof of the theorem.

LEMMA 20.6. *If $k \geqslant 2l$, then*

$$\sum_{s=0}^{l} \frac{(-1)^s(k-2s+n-1)(k-l-s+n-2)!}{s!\,(l-s)!\,(k+n-s-1)!} = \begin{cases} 0, & \text{if } l>0, \\ 1, & \text{if } l=0, \end{cases}$$

and

$$\sum_{s=0}^{l} \frac{(-1)^s(k-2s+n-1)(k-l-s+n-2)!s}{s!\,(l-s)!\,(k+n-s-1)!}$$

$$= \begin{cases} \dfrac{(-1)^l(n+k-2l-1)!\,(2l-2)!}{(n+k-2)!l!\,(l-1)!}, & \text{if } l>0, \\ 0, \quad \text{if } l=0. \end{cases}$$

PROOF. Let us compute for instance the first sum, which we denote by S. We have

$$\begin{aligned} S &= \frac{(-1)^l(k-2l+n-1)(k-2l+n-2)!}{l!\,(k+n-l-1)!} \\ &\quad + (-1)^{l-1}(k-2l+n+1)(k-2l+n-1)! + \cdots \\ &= \frac{(-1)^l(k-2l+n-1)!}{(l-1)!\,(k+n-l-1)!}\left[-\frac{1}{l}+\frac{k-2l+n+1}{k+n-l}\right] + \cdots \\ &= \frac{(-1)^{l-1}(k-2l+n-1)!}{l!\,(k+n-l)!}(l-1)(k+n-2l) + \cdots \\ &= \frac{(-1)^{l-1}(k-2l+n)!\,(l-1)}{l!\,(k+n-l)!} \\ &\quad + \frac{(-1)^{l+2}(k-2l+n+3)(k-2l+n)!}{(l-2)!2!(k+n-l+1)!} + \cdots \\ &= \frac{(-1)^{l+2}(k-2l+n+1)!(l-2)}{2!l(l-2)!(k+n-l+1)!} + \cdots \\ &= -\frac{(k-l+n-2)!}{(l-1)!l!(k+n-2)!} + \frac{(k+n-1)(k-l+n-2)!}{l!(k+n-1)!} = 0. \end{aligned}$$

The second sum can be computed in the same way.

COROLLARY 20.7. *Let* $P = \sum_0^m P_k$ *be an arbitrary polynomial, where the* P_k *are homogeneous polynomials of degree* k. *Then* P *can be extended holomorphically from* $\partial B(0, r)$ *into* $B(0, r)$ *if and only if*

$$\sum_{k=0}^{m} \mathfrak{L}_k p_k \big|_{\partial B(0, r)} = 0.$$

We note that the condition $\mathfrak{L}_k P_k = 0$ on $\partial B(0, r)$ may be replaced by the condition $\mathfrak{L}'_k P_k = 0$ in B, where

$$\mathfrak{L}'_k = \sum_{j=1}^{n} \bar{z}_j \frac{\partial}{\partial \bar{z}_j} - \sum_{l \geqslant 0} \frac{(n + k - 2l - 3)!\,(2l - 1)!!2^l}{(n + k - 2)!\,(l + 1)!} |z|^{2l+2} \Delta^{l+1}.$$

CHAPTER VI

FORMS ORTHOGONAL TO HOLOMORPHIC FORMS. WEIGHTED FORMULA FOR SOLVING THE $\bar{\partial}$-EQUATION, AND APPLICATIONS

§21. Forms orthogonal to holomorphic forms

In this section, we sharpen the results of §6. First some ancillary results:

LEMMA 21.1. *Let D be a bounded strictly pseudoconvex domain in $\mathbf{C}^n$. Then there exist positive constants r, δ and c such that:*

I) *For any ζ with $|\rho(\zeta)|< r$ in $B(\zeta, \delta)$, we can perform a smooth change of variables $\tau = \tau(z)$ with the properties*

1) $\tau_1 = \rho(z) - \rho(\zeta)$; $\tau_2 = \operatorname{Im} \Phi(\zeta, z)$;

2) $|\zeta - z|/c \geqslant |\tau(z)| \geqslant c|\zeta - z|$; $z \in B(\zeta, \delta)$;

3) $1/c \geqslant |\partial\tau/\partial z| \geqslant c$; $z \in B(\zeta, \delta)$.

II) *For any z such that $|\rho(z)|< r$ in $B(z, \delta)$, we can perform a smooth change of variables $\tau = \tau(\zeta)$ with the properties*

1) $\tau_1 = \rho(\zeta) - \rho(z)$; $\tau_2 = \operatorname{Im} \Phi(\zeta, z)$;

2) $|\zeta - z|/c \geqslant |\tau(z)| \geqslant c|\zeta - z|$, $\zeta \in B(z, \delta)$;

3) $1/c \geqslant |\partial\tau/\partial\zeta| \geqslant c$; $\zeta \in B(z, \delta)$.

The function $\Phi(\zeta, z)$ was introduced in §4.2°.

PROOF. We need one more property of Khenkin's barrier function not noted in §4.2°: there exist $r > 0$ and a nowhere-vanishing continuous function h on $\{z: |\rho(z)|< r\}$ such that, for $z \in V$ with $|\rho(z)|< r$,

$$d_z\Phi(\zeta, z)\big|_{z=\zeta} = d_\zeta\Phi(\zeta, z)\big|_{z=\zeta} = h\,d\rho$$

(see Øvrelid [1]).

From this we conclude that the maps

$$\begin{cases} \tau_1 = \rho(z), \\ \tau_2 = \operatorname{Im} \Phi(\zeta, z), \end{cases} \quad \text{and} \quad \begin{cases} \tau_1 = \rho(\zeta), \\ \tau_2 = \operatorname{Im} \Phi(\zeta, z), \end{cases}$$

have rank 2 at points where $z = \zeta$ and grad $\rho(z) \neq 0$. The lemma follows easily from this.

LEMMA 21.2. *Let*

$$I_{\alpha,k,l}(\varepsilon) = \int_{\substack{|\tau|<1 \\ \tau_1+\varepsilon>0}} \frac{(\tau_1 + \varepsilon)^\alpha d\tau_1 \cdots d\tau_{2n}}{(|\tau_1| + |\tau_2| + \varepsilon + |\tau|^2)^k |\tau|^l},$$

where α, k and l are real and $\alpha > -1$. Then
a) $I_{\alpha,k,2n-1}(\varepsilon) = O(1)\cdot\varepsilon^{1-k+\alpha}$; $k - \alpha > 1$;
b) $I_{\alpha,k,2n-3-2m}(\varepsilon) = O(1)\varepsilon^{-k+\alpha+m+5/2}$; $k - \alpha - m > 5/2$; $m > 0$.

PROOF. In the integral defining $I_{\alpha,k,l}$, we introduce spherical coordinates $\tau_1 = r\cos\varphi_1$, $\tau_2 = r\sin\varphi_1\cos\varphi_2, \ldots,$ and perform the change of variables $s_j = \cos\varphi_j$, $j = 1, 2$. Then

$$I_{\alpha,k,l}(\varepsilon) = O(1)\int_{\substack{0<r,\, s_2<1 \\ |s_1|<1 \\ rs_1+\varepsilon>0}} \frac{(rs_1 + \varepsilon)^\alpha r^{2n-1-l} dr ds_1 ds_2}{\left[r\left(|s_1| + s_2(1 - s_1^2)^{1/2}\right) + \varepsilon + r^2\right]^k}.$$

We use the inequality $|s_1| + s_2(1 - s_1^2)^{1/2} \leqslant \frac{1}{4}(|s_1| + s_2)$ for $|s_1| \leqslant 1$ and $0 \leqslant s_2 \leqslant 1$. Setting $s_1 = su$ and $s_2 = s(1 - |u|)$, we have

$$I_{\alpha,k,l}(\varepsilon) = O(1)\int_{\substack{0<r,\, s<2 \\ |u|<1 \\ rsu+\varepsilon>0}} \frac{(rsu + \varepsilon)^\alpha r^{2n-1-l} s dr ds du}{(rs + \varepsilon + r^2)^k}.$$

For $l = n - 1$, we go over to the coordinates (v, s, u), where $v = \varepsilon^{-1}rs$. Since $rs + \varepsilon + r^2 > rs + \varepsilon$, we get

$$I_{\alpha,k,2n-1}(\varepsilon) = O(1) \cdot \varepsilon^{1-k+\alpha}\left[\int_0^1 \frac{dv}{(v + 1)^k}\int_{-1}^1 (vu + 1)^\alpha du + \int_1^{4\varepsilon^{-1}} \frac{dv}{(v + 1)^k}\int_{-v^{-1}}^1 (vu + 1)^\alpha du\right].$$

Assertion a) follows.

In the case b), we have

$$I_{\alpha,k,2n-3-2m}(\varepsilon) = O(1)\int_0^2 r^{2m} dr \int_0^2 \frac{rds}{(rs + \varepsilon + r^2)^k}\int_{-\varepsilon(rs)^{-1}}^1 (rsu + \varepsilon)^\alpha rsdu$$

$$= O(1)\int_0^2 r^{2m} dr \int_0^2 \frac{rds}{(rs + \varepsilon + r^2)^{k-\alpha-1}} = O(1)\int_0^2 \frac{r^{2m} dr}{(\varepsilon + r^2)^{k+\alpha-2}}$$

$$= O(1)\int_0^2 \frac{dr}{\left(\sqrt{\varepsilon} + r\right)^{2(k-\alpha-m-2)}}.$$

Assertion b) now follows by direct integration.

LEMMA 21.3. *Let $D \subset \mathbf{R}^n$ be a bounded domain with smooth boundary and let $f \in C^1(D)$. Suppose* $|\operatorname{grad} f(z)| = O(1)|\rho(z)|^{\lambda-1}$ *for* $0<\lambda<1$. *Then* $f \in C^{0,\lambda}(\bar{D})$.

This assertion is a slight generalization of a result of Hardy and Littlewood (see, for example, Goluzin [1], Chapter IX, §5, Theorem 3).

We now improve the estimate for the solution of the $\bar{\partial}$-problem given by Theorem 4.4.

THEOREM 21.4. *Let D be a bounded strictly pseudoconvex domain in $\mathbf{C}^n$ with boundary of class C^{m+2}, and let $\gamma \in Z^{m,\lambda}_{(p,q)}(\bar{D})$, $m \geqslant 0$, $\lambda > 0$. Then there exists a form α satisfying $\bar{\partial}\alpha = \gamma$ in D, with*

$$\begin{aligned} &\alpha \in C^{m,\lambda'+1/2}_{(p,q-1)}(\bar{D}), && 0<\lambda'<\lambda, \text{ if } 0<\lambda\leqslant \tfrac{1}{2};\\ &\alpha \in C^{m+1,\lambda'-1/2}_{(p,q-1)}(\bar{D}), && \tfrac{1}{2}<\lambda'<\lambda, \text{ if } \tfrac{1}{2}<\lambda\leqslant 1. \end{aligned} \tag{21.1}$$

PROOF. We proceed as in Theorem 4.4. We need only show that the form

$$\int_{(V\setminus D)_\varepsilon} \bar{\partial}\tilde{\gamma}(\zeta) \wedge \mu^1_{q-1} \wedge D_{p,n-p}(\partial z, \partial\zeta)$$

satisfies (21.1). Lemmas 1.4 and 4.1 show that it is enough to verify that (21.1) is true for the function

$$H(z) = \int_{(V\setminus D)_\zeta} \frac{h(\zeta)\varphi(\zeta,z)(\bar{\zeta}_k - \bar{z}_k)}{\Phi^{r+1}(\zeta,z)|\zeta - z|^{2(n-r-1)}} d\bar{\zeta}\wedge d\zeta, \tag{21.2}$$

with $h \in C^{m-1,\lambda}(\mathbf{C}^n)$ and $h \equiv 0$ on D. If $m=0$, we may suppose that $h \in C(V\setminus D)$ and $|h(\zeta)| = O(1)|\rho(\zeta)|^{\lambda-1}$; for this, we have to extend γ to all of $\mathbf{C}^n$ by means of Theorem 3.2 and Supplement 3.3 of Malgrange [1], Chapter 1.

For arbitrary m, by working out the argument in Lemma 4.2 more carefully, we see that $h(\zeta) = O(1)|\rho(\zeta)|^{m-1+\lambda}$. We shall carry out the rest of the proof in the case $\lambda > \frac{1}{2}$. The case $\lambda \leqslant \frac{1}{2}$ can be handled similarly.

Differentiating under the integral sign and using the argument in Lemma 4.2, we conclude that, for $z \in D$,

$$|D^{I,J}H(z)| = O(1)\sum_{j=0}^{2} \int_{(V\setminus D)_\zeta} \frac{\rho^{\lambda-1}(\zeta)\,dv_\zeta}{|\Phi(\zeta,z)|^{3-j}|\zeta - z|^{2n-3+j}},$$
$$\|I\| + \|J\| = m+2.$$

Choose r and δ according to Lemma 21.1 and split the integrals over $V\setminus D$ into integrals over $(V\setminus D)\setminus B(z,\delta)$ and $(V\setminus D)\cap B(z,\delta)$. The first integrals can be estimated above by a constant independent of z. Thus

$$|D^{I,J}H(z)| = O(1)\left[1 + \sum_{j=0}^{2} \int_{[(V\setminus D)\cap B(z,\delta)]_\zeta} \frac{\rho^{\lambda-1}(\zeta)\,dv_\zeta}{|\Phi(\zeta,z)|^{3-j}|\zeta - z|^{2n-3+j}}\right]$$

$$= O(1)\left[1 + \sum_{j=0}^{2}\right.$$

$$\left.\cdot\int_{[(V\setminus D)\cap B(z,\delta)]_\zeta} \frac{\rho^{\lambda-1}\,dv_\zeta}{\left(c_1(\rho(\zeta) - \rho(z)) + c_2|\zeta - z|^2 + |I_m\Phi(\zeta,z)|\right)^{3-j}|\zeta - z|^{2n-3+j}}\right].$$

Here, we have used inequality 4 from §4.2°. Now make the change of variables occurring in part II of Lemma 21.1. Using the definition of the $I_{\alpha,k,l}$, the obvious inequalities

$$I_{\lambda-1,2,2n-2}(\varepsilon) = O(1)I_{\lambda-1,1\frac{1}{2},2n-1}(\varepsilon),$$
$$I_{\lambda-1,1,2n-1}(\varepsilon) = O(1)I_{\lambda-1,1\frac{1}{2},2n-1}(\varepsilon)$$

and Lemma 21.2, we get for $\|I\| + \|J\| = m + 2$

$$|D^{I,J}H(z)| = O(1)\left[1 + \sum_{j=0}^{2} I_{\lambda-1,3-j,2n-3+j}(|\rho(z)|)\right]$$
$$= O(1)\left[1 + I_{\lambda-1,3,2n-3}(|\rho(z)|) + I_{\lambda-1,1\frac{1}{2},2n-1}(|\rho(z)|)\right]$$
$$= O(1)|\rho(z)|^{\lambda-1\frac{1}{2}}.$$

Hence we conclude by Lemma 21.3 that all derivatives of H of order $m + 1$ lie in $C^{0,\lambda-1/2}(D)$. Thus H satisfies (21.1).

We next give the sharpened version of Theorem 6.1.

THEOREM 21.5. *Let D be a bounded strictly pseudoconvex domain in $\mathbf{C}^n$, $n > 1$, with $\partial D \in C^{m+2}$, $m \geq 0$, and let $\alpha \in C^{m,\lambda}_{(p,n-1)}(\partial D)$, $\lambda > 0$. Then $\alpha \in A^{\perp}_{n-p}(\overline{D})$ if and only if there exists a form γ satisfying* (21.1) *such that $\bar\partial\gamma|_{\partial D} = \alpha$.*

PROOF. The sufficiency was proved in §6. We prove *necessity*. By Theorem 2.12,

$$\alpha = \alpha_1|_{\partial D} - \partial_2|_{\partial D}, \tag{21.3}$$

where $\alpha_1 \in Z^{m,\lambda'}_{(p,n-1)}(\overline{D})$ and $\alpha_2 \in Z^{m,\lambda'}_{(p,n-1)}(CD)$. By Theorem 21.4, $\alpha_1 = \bar\partial\gamma_1$ and γ_1 satisfies (21.1). Extend α_2 to a form $\tilde\alpha_2 \in C^{m,\lambda'}_{(p,n-1)}(\mathbf{C}^n)$. By (1.1), for

$z \in (V \cap G) \setminus \bar{D}$ we have

$$\alpha_2 = \tilde{\alpha}_2 = I^1_{p,n-1}(G, \tilde{\alpha}_2) - I^2_{p,n-1}(G, \bar{\partial}\tilde{\alpha}_2) - \bar{\partial} I^2_{p,n-2}(G, \tilde{\alpha}_2). \quad (21.4)$$

Here G is any domain of holomorphy with boundary of class C^{m+2} such that $D \Subset G$. Using Lemma 1.3 and the fact that $\bar{\partial}\tilde{\alpha}_2 = \bar{\partial}\alpha_2 = 0$ on $C\bar{D}$, we get for $z \in G \setminus D$

$$I^2_{p,n-1}(G, \bar{\partial}\tilde{\alpha}_2)(\zeta) = I^2_{(p,n-1)}(D, \bar{\partial}\tilde{\alpha}_2)(\zeta)$$
$$- \bar{\partial}_\zeta \int_{D_z} \bar{\partial}\tilde{\alpha}_2 \wedge \mu_0 - \int_{D_z} \bar{\partial}\tilde{\alpha}_2 \wedge W_{p,0}(u, \zeta, z). \quad (21.5)$$

Here μ_0 is the form defined in Lemma 1.3, for the choice

$$w^1 = t = (\bar{\zeta} - \bar{z})/|\zeta - z|^2 \quad \text{and} \quad w^2 = u = P(\zeta, z)/\Phi(\zeta, z).$$

The coefficients of $W_{p,0}(u, \zeta, z)$ are holomorphic in z, and

$$\tilde{\alpha}_2 |_{\partial D} = \alpha_2 |_{\partial D} = \alpha_1 |_{\partial D} - \alpha \in A^{\perp}_{n-p}(\bar{D}),$$

so that the last integral on the right side of (21.5) vanishes by Stokes' theorem. The integral

$$\gamma_2(\zeta) = \int_{D_z} \bar{\partial}\tilde{\alpha}_2 \wedge \mu_0$$

and its derivatives can be estimated as in the proof of Theorem 21.4 to conclude that γ_2 satisfies (21.1) with $G \setminus \bar{D}$ in the place of D. Extend γ_2 to all of G with the same smoothness, and denote the extension also by γ_2. Next, since $I^1_{p,n-1}(G, \tilde{\alpha}_2) \in Z^{\infty}_{(p,n-1)}(G)$, we have $\bar{\partial}\gamma_3 = I^1_{p,n-1}(G, \tilde{\alpha}_2)$, where $\gamma_3 \in C^{\infty}_{(p,n-2)}(G)$. Then it follows from (21.3)–(21.5) that $\alpha = \bar{\partial}\gamma |_{\partial D}$, where $\gamma = \gamma_1 + \gamma_2 - \gamma_3 + I^2_{p,n-2}(G, \tilde{\alpha}_2)$. Corollary 2.4 now finishes the proof.

If we use Theorem 21.5 in place of Theorem 6.1 in the proofs of Theorems 13.2 and 13.4, then we can weaken their hypotheses: instead of the condition $\mu_z \in C^{1,\lambda}_{2n-1}(\partial D)$, $\lambda > 0$, we need only require $\mu_z \in C^{0,\lambda}_{2n-1}(\partial D)$, $\lambda > \frac{1}{2}$.

We state the next theorem without proof.

THEOREM 21.6. *Let D be a bounded strictly pseudoconvex domain in $\mathbf{C}^n$, and suppose α is a $(p, n-1)$-form on ∂D with coefficients in $L^1(\partial D)$. Then $\alpha \in A^{\perp}_{n-1}(\bar{D})$ if and only if there exists $\beta \in Z_{p,n-1}(D)$ such that*

$$\lim_{\varepsilon \to 0} \int_{\{\rho(z) = -\varepsilon\}} \beta(z) \wedge \varphi(z) = \int_{\partial D} \alpha(z) \wedge \varphi(z) \quad (21.6)$$

for any $\varphi \in C_{(n-p,0)}(\bar{D})$. Also,

$$\int_{\{\rho(z) = -\varepsilon\}} \left(|\beta(z) \wedge \bar{\partial}\rho(z)| + |\rho(z)|^{1/2} |\beta(z)| \right) dv_z = O(1) \|\alpha\|_{L^1}. \quad (21.7)$$

Here, as before, ρ is a function defining D; by the absolute value of a form we mean the sum of the absolute values of its coefficients, and $\|\cdot\|_{L^1}$ denotes the L^1 norm of the absolute value of the form. (21.6) says that, in some (generalized) sense, the restriction of β to ∂D coincides with α. (21.7) says that the tangential part of β has the same properties as α.

For forms of type $(p, n-1)$, the $\bar\partial$-problem is solvable not only for domains of holomorphy but also for domains D with $\partial D \in C^2$ for which the restriction of the Levi form (cf. §4.2°) to the complex tangent space to ∂D at a point ζ has at least one positive eigenvalue for each $\zeta \in \partial D$. In the C^∞ case, this was done by Kohn and Rossi [1]. By combining the methods of proof of Fischer and Lieb [1] with those of Theorem 21.4, it is possible to show that Theorem 21.4 is valid for $q = n-1$ and for domains of the above kind. After making the corresponding changes in the proof of Theorem 6.2, we thus obtain the following improved version:

THEOREM 21.7. *Let $D = \Omega \setminus \bigcup_1^k \Omega_j$, where Ω is a strictly pseudoconvex domain with $\partial\Omega \in C^{m+2}$, $m \geq 1$, and the $\Omega_j \Subset \Omega$ are domains with boundaries of class C^{m+2} such that $\bar\Omega_j \cap \bar\Omega_l = \varnothing$ for $j \neq l$, $j, l = 1,\ldots,k$, and the restriction of the Levi form to the complex tangent space to $\partial\Omega_j$ at ζ has at least one positive eigenvalue at each $\zeta \in \partial\Omega_j$, $j = 1,\ldots,k$. Let $\alpha \in C^{m,\lambda}_{(p,n-1)}(\partial D)$. Then $\alpha \in A^{\perp}_{n-p}(\bar D)$ if and only if there exists a form $\tilde\alpha$ such that $\tilde\alpha|_{\partial D} = \alpha$ and*

$$\tilde\alpha \in Z^{m-1,\lambda'+1/2}_{(p,n-1)}(\bar D), \qquad 0 < \lambda' < \lambda, \textit{ if } 0 < \lambda \leq 1/2;$$
$$\tilde\alpha \in Z^{m,\lambda'-1/2}_{(p,n-1)}(D), \qquad 1/2 < \lambda' < \lambda, \textit{ if } 1/2 < \lambda \leq 1.$$

§22. Generalization of Theorem 8.1

THEOREM 22.1. *Let D be a bounded domain in $\mathbf{C}^n$ with boundary of class C^2, and suppose that for each $\alpha \in A^{\perp}(\bar D) \cap C^{\infty}_{(n,n-1)}(\bar D)$ there is an $\tilde\alpha \in Z^{1}_{(n,n-1)}(\bar D)$ such that $\tilde\alpha|_{\partial D} = \alpha$. Then $D = \Omega \setminus \bigcup_1^k \bar\Omega_j$, where the $\Omega_j \Subset \Omega$ are domains such that $\bar\Omega_j \cap \bar\Omega_l = \varnothing$ for $j \neq l$, $j, l = 1,\ldots,k$, and the restriction of the Levi form of $\partial\Omega_j$ to the complex tangent space to $\partial\Omega_j$ at ζ has at least one nonnegative eigenvalue at each $\zeta \in \partial\Omega_j$, $j = 1,\ldots,k$.*

Theorems 21.7 and 22.1 give similar necessary and sufficient conditions on the bounded components of the complement of D in order that all forms in $A^{\perp}_{n-p}(D)$ have $\bar\partial$-closed extensions into D. However, the most interesting part of the boundary, namely that of the unbounded component of the complement, remains unexplored.

PROOF. Suppose that, for some j, Ω_j does not satisfy the conclusion of the theorem, i.e. there exists $\zeta^0 \in \partial\Omega_j$ at which all the eigenvalues of the Levi form are negative. Then there exists a neighborhood U of ζ^0 such that for any $\zeta \in U \cap \overline{\Omega}_j$ the function

$$F_\zeta(z) = \sum_{l=1}^{n} \frac{\partial\rho}{\partial z_l}(\zeta)(z_l - \zeta_l) + \sum_{l,m=1}^{n} \frac{\partial^2\rho}{\partial z_l \partial z_m}(\zeta)(z_l - \zeta_l)(z_m - \zeta_m)$$

vanishes in $U \cap C\Omega_j$ if and only if $z = \zeta \in \partial\Omega_j \cup U$ (see, for example, Gunning and Rossi [1], Chapter IX, §B). A similar assertion was used in the first part of the proof of Theorem 8.1. It follows that there exist $\zeta^1 \in U \cap \Omega_j$, $z^0 \in U \cap C\overline{\Omega}_j$ and $r > 0$ such that the function $f(z) = 1/F_{\zeta^1}(z)$ satisfies

$$|f(z^0)| > |f(z)|, \qquad z \in C\Omega_j \cap \partial B(\zeta^0, r), \tag{22.1}$$

and $C\overline{\Omega}_j \cap B(\zeta^0, r)$ has piecewise smooth boundary. The form

$$\alpha \begin{cases} U_{0,0}(\zeta, z^0), & \zeta \in \partial\Omega_j, \\ 0, & \zeta \in \partial D \setminus \partial\Omega_j, \end{cases}$$

lies in $A^\perp(\overline{D}) \cap C^\infty_{(n,n-1)}(\overline{D})$; hence there exists $\tilde{\alpha} \in Z^1_{(n,n-1)}(\overline{D})$ such that $\tilde{\alpha}|_{\partial D} = \alpha$. As in the proof of Theorem 8.1, we see that $U_{0,0}(\zeta, z^0) = \bar{\partial}\gamma$ in Ω_j, where

$$\gamma(\zeta) = -\int_{(\Omega_j)_w} U_{0,0}(w, z^0) U_{n,n-2}(w, \zeta) + \int_{D_w} \tilde{\alpha}(w) \wedge U_{n,n-2}(w, \zeta).$$

By Corollary 2.4, $\gamma \in C^1_{(n,n-2)}(\Omega_j)$. By the Martinelli-Bochner formula (i.e. (1.1) for $p = q = 0$) and Stokes' theorem, for any $g \in A\big(C\Omega_j \cap B(\zeta^0, r)\big)$ we have

$$\begin{aligned} g(z^0) &= \int_{\partial(C\Omega_j \cap B(\zeta^0, r))_\zeta} g(\zeta) U_{0,0}(\zeta, z^0) \\ &= \int_\Gamma g(\zeta) \wedge \gamma(\zeta) + \int_{(C\Omega_j \cap \partial B(\zeta^0, r))_\zeta}. \end{aligned}$$

Here $\Gamma = \partial\Omega_j \cap \partial B(\zeta^0, r)$. It follows that, for some constant c,

$$|g(z^0)| \leqslant C \max_{\zeta \in C\Omega_j \cap \partial B(\zeta^0, r)} |g(z)|. \tag{22.2}$$

But it is clear from (22.1) that if $g(z) = f^m(z)$ with m sufficiently large, then (22.2) cannot hold. This contradiction proves the theorem.

§23. Weighted formula for solving the $\bar\partial$-problem in strictly convex domains and zeros of functions of the Nevanlinna-Dzhrbashyan class

To obtain certain precise results in strictly pseudoconvex domains, it became necessary to have the solutions of the $\bar\partial$-equation with a right-hand side that grows near the boundary expressed by a formula. For simplicity, we present the formula here only for a strictly convex domain.

Let D be a strictly convex domain in $\mathbf{C}^n$, i.e. a domain of the form $D = \{z \in U, \rho(z) < 0\}$, where U is a neighborhood of $\bar{D}$, $\rho \in C^2(U)$, and the second differential (i.e. Hessian) of ρ is positive definite at all $\zeta \in U$. Hence $\operatorname{grad} \rho \,|_{\partial D} \neq 0$.

For a strictly convex domain, the Khenkin barrier function (see §4.2°) may be introduced as follows:

$$P(\zeta) = (P_1, \dots, P_n) = \left(\frac{\partial\rho}{\partial\zeta_1}, \dots, \frac{\partial\rho}{\partial\zeta_n}\right); \qquad \Phi(\zeta, z) = \langle P, \zeta - z\rangle.$$

We set $\tilde\Phi(\zeta, z) = \Phi(\zeta, z) - \rho(\zeta)$. By Taylor's theorem and the strict convexity of ρ, we have

$$\rho(z) \geqslant \rho(\zeta) - 2\operatorname{Re}\Phi(\zeta, z) + c_1|\zeta - z|^2$$

for $z, \zeta \in D'$, $D \Subset D' \Subset U$, with a constant $c_1 > 0$ depending only on D'. Therefore

$$\begin{aligned} 2\operatorname{Re}\Phi(\zeta, z) &\geqslant \rho(\zeta) - \rho(z) + c_1|\zeta - z|^2, \\ 2\operatorname{Re}\tilde\Phi(\zeta, z) &\geqslant -\rho(\zeta) - \rho(z) + c_1|\zeta - z|^2. \end{aligned} \tag{23.1}$$

Hence

$$|\Phi(\zeta, z)| = O(1)|\tilde\Phi(\zeta, z)|, \tag{23.2}$$

$$|\rho(\zeta) - \rho(z)| - \rho(\zeta) - \rho(z) + |\operatorname{Im}\Phi(\zeta, z)| + |\zeta - z|^2 = O(1)|\Phi(\zeta, z)|. \tag{23.3}$$

The constants in $O(1)$ depend only on D' and not on $z, \zeta \in D'$.

For a (p, q)-form γ, we define

$$\begin{aligned} (G^k_{p,q}\gamma)(z) = {} & -I^2_{p,q-1}(D, \alpha_k(\zeta, z)\gamma(\zeta))(z) \\ & - \int_{D_\zeta} \bar\partial_\zeta \alpha_k(\zeta, z) \wedge \gamma(\zeta) \wedge \mu^1_{q-1} \wedge \frac{1}{p!\,(n-p)!} D_{p,n-p}(\partial z, \partial\zeta). \end{aligned} \tag{23.4}$$

Here, μ^1_{q-1} is given by Lemma 1.4, with $w = t = (\bar\zeta - \bar z)/|\zeta - z|^2$ and $u = P(\zeta)/\Phi(\zeta, z)$, and α_k is defined by

$$\alpha_k(\zeta, z) = \left[1 - \left(\frac{\Phi(\zeta, z)}{\tilde\Phi(\zeta, z)}\right)^{2n}\right]^k = \left(-\frac{\rho(\zeta)}{\tilde\Phi(\zeta, z)}\right)^k \left[\sum_{j=0}^{2n-1} \left(\frac{\Phi(\zeta, z)}{\tilde\Phi(\zeta, z)}\right)^j\right]^k. \tag{23.5}$$

We have

$$\bar\partial_\zeta \alpha_k(\zeta, z) = -k2n\alpha_{k-1}(\zeta, z)\left(\frac{\Phi(\zeta, z)}{\tilde\Phi(\zeta, z)}\right)^{2n-1}\left[\frac{\Phi(\zeta, z)\bar\partial\rho(\zeta) - \rho(\zeta)\bar\partial_\zeta\Phi(\zeta, z)}{\tilde\Phi^2(\zeta, z)}\right]. \tag{23.6}$$

From Lemma 4.1 and (23.6), and the fact that μ^1_{q-1} is a linear combination of forms of the type (1.10), we see that the kernels of the second term on the right in (23.4) have no singularitites. Also, α_k contains a factor $\rho^k(\zeta)$; hence the operator $G^k_{p,q}$ can be applied to forms γ such that $\rho^{k-1}(\zeta)\gamma(\zeta) \in L^1_{(p,q)}(D)$.

THEOREM 23.1. *Let $q > 0$, and suppose $\gamma \in Z^1_{(p,q)}(D)$ and $\theta > 0$ are such that $\rho^\theta\gamma \in L^1_{(p,q)}(D)$ and $\rho^{\theta-1/2}\bar\partial\rho \wedge \gamma \in L^1_{(p,q+1)}(D)$. Then for all $k \geqslant \theta + 1$ the forms $\beta_k = G^k_{p,q}\gamma$ satisfy $\bar\partial\beta_k = \gamma$ in D, and $\rho^{\theta-1}\beta_k \in L^1_{p,q-1}(D)$.*

PROOF. Assume first that $\gamma \in Z^1_{p,q}(\bar D)$. Substitute $\alpha_k(\zeta, z)\gamma(\zeta)$ in the Martinelli-Bochner-Koppelman formula (1.1). Since $\alpha_k(\zeta, z) = 0$ for $\zeta \in \partial D$, α_k is holomorphic in z on D and $\bar\partial\gamma = 0$, we get

$$\gamma(z) = -I^2_{p,q}\big(D, \bar\partial_\zeta\alpha_k(\zeta, z) \wedge \gamma\big) - \bar\partial I^2_{p,q-1}\big(D, \alpha_k(\zeta, z)\gamma\big). \tag{23.7}$$

We transform the first term on the right above by Lemma 1.4 to

$$\begin{aligned} I^2_{p,q}\big(D, \bar\partial_\zeta\alpha_k(\zeta, z) \wedge \gamma\big) &= \int_{D_\zeta} \bar\partial_\zeta\alpha_k(\zeta, z) \wedge \gamma(\zeta) \wedge U_{p,q}(\zeta, z) \\ &= \int_{D_\zeta} \bar\partial_\zeta\alpha_k(\zeta, z) \wedge \gamma(\zeta) \wedge \left(\bar\partial_z\mu^1_{q-1} + \bar\partial_\zeta\mu^2_q\right) \wedge \frac{1}{p!(n-p)!}D_{p,n-p}(\partial z, \partial\zeta) \\ &= \bar\partial_z\int_{D_\zeta} \bar\partial_\zeta\alpha_k(\zeta, z) \wedge \gamma(\zeta) \wedge \mu^1_{q-1} \wedge \frac{1}{p!(n-p)!}D_{p,n-p}(\partial z, \partial\zeta) \\ &\quad + (-1)^{p+q-1}\int_{D_\zeta} d\left[\bar\partial_\zeta\alpha_k(\zeta, z) \wedge \gamma(\zeta) \wedge \mu^2_q \wedge \frac{1}{p!(n-p)!}D_{p,n-p}(\partial z, \partial\zeta)\right]. \end{aligned}$$

As already observed, the integrals have no singularities, so it is legitimate to bring $\bar\partial_z$ outside the integral sign. Now, since $\bar\partial_\zeta\alpha_k(\zeta, z) = 0$ for $\zeta \in \partial D$ and $k > 1$, we see by Stokes' theorem that the second integral above vanishes.

Hence (23.7) yields $\gamma = \bar\partial G^k_{p,q}\gamma$. In the general case, we apply this equality to the form $\gamma_r(z) = \gamma(rz) \in Z^1_{p,q}(\bar D)$ and pass to the limit as $r \to 1$.

We proceed to estimate the solution thus obtained. From (23.6) and the definition of $G^k_{p,q}$, we have

$$\int_{D_z} |\rho^{\theta-1}(z)| \left|(G^k_{p,q}\gamma)(z)\right| dv_z = O(1)\int_{D_\zeta} |\gamma(\zeta)| dv_\zeta \int_{D_z} |\rho^{\theta-1}(z)|$$

$$\cdot \Bigg\{ \left|U_{p,q-1}(\zeta, z)\right| |\alpha_k(\zeta, z)|$$

$$+ |\alpha_{k-1}(\zeta, z)| \frac{|\Phi(\zeta, z)|^{2n-1}}{|\tilde\Phi(\zeta, z)|^{2n+1}} |\rho(\zeta)| \left|\bar\partial_\zeta \Phi(\zeta, z)\right| \left|\mu^1_{q-1}\right| \Bigg\} dv_z$$

$$+ O(1)\int_{D_\zeta} |\gamma(\zeta) \wedge \bar\partial\rho(\zeta)| dv_\zeta \int_{D_z} |\rho^{\theta-1}(z)| \, |\alpha_{k-1}(\zeta, z)| \frac{|\Phi(\zeta, z)|^{2n}}{|\tilde\Phi(\zeta, z)|^{2n+1}} \left|\mu^1_{q-1}\right| dv_z .$$

Here, as before, dv denotes the volume element, and the absolute value of a form is the sum of the absolute values of its coefficients.

Next, using (1.4), Lemma 4.1, (23.5), (23.2), (23.3), the relation

$$\left|\bar\partial_\zeta \Phi(\zeta, z)\right| = \left|\langle \bar\partial P(\zeta), \zeta - z\rangle\right| = O(1)|\zeta - z|,$$

and the definition of μ^1_{q-1} (see Lemma 1.4), we get

$$\int_{D_z} |\rho^{\theta-1}(z)| \left|(G^k_{p,q}\gamma)(z)\right| dv_z = O(1)\int_{D_\zeta} |\gamma(\zeta)| \, |\rho^k(\zeta)| dv_\zeta \int_{D_z} |\rho^{\theta-1}(z)|$$

$$\times \left\{ \frac{1}{|\tilde\Phi(\zeta, z)|^k |\zeta - z|^{2n-1}} + \frac{1}{|\tilde\Phi(\zeta, z)|^{k+2} |\zeta - z|^{2n-4}} \right\} |\rho^{\theta-1}(z)| dv_z$$

$$+ O(1)\int_{D_\zeta} |\gamma(\zeta) \wedge \bar\partial\rho(\zeta)| \, |\rho^{k-1}(\zeta)| dv_\zeta \int_{D_z} \frac{|\rho^{\theta-1}(z)| \, dv_z}{|\tilde\Phi(\zeta, z)|^{k+1} |\zeta - z|^{2n-3}} .$$

Hence, by the same argument as in the proof of Theorem 21.4,

$$\int_{D_z} |\rho^{\theta-1}(z)| \left|(G^k_{p,q}\gamma)(z)\right| dv_z$$

$$= O(1)\Bigg\{ 1 + \int_{D_\zeta} |\gamma(\zeta)| \, |\rho^k(\zeta)| dv_\zeta \left[I_{\theta-1,k,2n-1}(|\rho(\zeta)|) + I_{\theta-1,k+2,2n-4}(|\rho(\zeta)|) \right]$$

$$+ \int_{D_\zeta} |\gamma(\zeta) \wedge \bar\partial\rho(\zeta)| \, |\rho^{k-1}(\zeta)| dv_\zeta I_{\theta-1,k+1,2n-3}(|\rho(\zeta)|) \Bigg\}.$$

An application of Lemma 21.2 finishes the proof of the theorem.

We state without proof an application of Theorem 23.1.

Let $D = \{z: \rho(z) < 0\}$ be a bounded domain in $\mathbf{C}^n$ with C^2 boundary. We denote by $N_\alpha(D)$ ($\alpha > 0$) the class of holomorphic functions F in D such that

$$\int_D |\rho(z)|^{\alpha-1} \ln^+|F(z)| d\sigma_{2n}(z) < \infty$$

(the Nevanlinna-Dzhrbashyan class). Here $d\sigma_k$ is the k-dimensional Lebesgue-Hausdorff measure, and $\ln^+ t = \max\{\ln t, 0\}$, $t > 0$.

THEOREM 23.2. *Let D be a strictly convex domain in $\mathbf{C}^n$. Then an $(n-1)$-dimensional analytic subset M of D is the set of zeros of a function from $N_\alpha(D)$, $\alpha \geqslant 0$, if and only if*

$$\int_M |\rho(z)|^{\alpha+1} d\sigma_{2n-2}(z) = \sum_j \gamma_j \int_{M_j} |\rho(z)|^{\alpha+1} d\sigma_{2n-2}(z) < \infty,$$

where the M_j are the irreducible components of M, and γ_j is the multiplicity of M_j in M.

CHAPTER VII

REPRESENTATION AND MULTIPLICATION OF DISTRIBUTIONS IN HIGHER DIMENSIONS

§24. Harmonic representation of distributions

Recently, there has been increased interest in the problem of multiplication of distributions, and there has been a lot of work on the problem. In the International Conference on Generalized Functions held in Moscow (November 1980), different approaches to solving this problem were discussed in several papers. Here we present a method based on representing the distributions by harmonic functions.

Let $\mathbf{R}^{n+1}$ be the space of the variables $(x_1,\dots,x_n, y) = (x, y)$, and let $\mathbf{R}^{n+1}_+ = \{(x, y) \in \mathbf{R}^{n+1}, y > 0\}$, then $\mathbf{R}^n = \{(x, y)\colon y = 0\}$. We shall say that a function f defined in $\mathbf{R}^{n+1}_+$ has *finite order of growth* as $y \to +0$, if for every compact set M in $\mathbf{R}^n$ there exist $c > 0$ and $m > 0$ such that $|f(x, y)| < cy^{-m}$ for $x \in M$ and $0 < y \leqslant 1$. Let H denote the set of harmonic functions f in $\mathbf{R}^{n+1}_+$ having finite order of growth as $y \to +0$, and K the set of harmonic functions in $\mathbf{R}^{n+1}_+$ which can be extended continuously to $\mathbf{R}^{n+1}_+ \cup \mathbf{R}^n$ and vanish on $\mathbf{R}^n$. Then $K \subset H$. Let $K(x, y)$ be the Poisson kernel for the half-space $\mathbf{R}^{n+1}_+$:

$$K(x, y) = \frac{\Gamma((n+1)/2)}{\pi^{(n+1)/2}} \cdot \frac{y}{\left(|x|^2 + y^2\right)^{(n+1)/2}}.$$

If $T \in E'(\mathbf{R}^n)$, then the harmonic function $T^*(x, y) = T_t K(x - t, y)$ is called the *Poisson representation* of T.

THEOREM 24.1. *If $T \in E'$, then $T^* \in H$ and represents T in the following sense*:

$$\lim_{\varepsilon \to +0} \int_{y=\varepsilon} T^*(x, y)\varphi(x)\,dx = \langle T, \varphi \rangle, \qquad \varphi \in \mathcal{O}_0(\mathbf{R}^n).$$

Here T^* extends continuously to $\mathbf{R}_+^{n+1} \cup (\mathbf{R}^n \setminus \operatorname{supp} T)$, and vanishes on $\mathbf{R}^n \setminus \operatorname{supp} T$, and $T^*(x, \varepsilon) = O(|x|^{-n-1})$ as $|x| \to \infty$. The Poisson representation of a derivative of T is the corresponding derivative of the Poisson representation of T.

This theorem generalizes Theorem 14.1, since (see §14) $\mathbf{C}^n = \mathbf{R}^{2n}$

$$\alpha_T(z)\big|_{y=\varepsilon} - \alpha_T(z)\big|_{y=-\varepsilon} = T^*(x_1, \ldots, x_{2n-1}, \varepsilon).$$

The proof of Theorem 24.1 is the same as that of Theorem 14.1. We need only show that $T^* \in H$. But since $T \in E'$, it is bounded with respect to some seminorm

$$\|\varphi\|_K^m = \sup_{x \in K,\, \|I\| \leqslant m} |D^I \varphi(x)|, \qquad K \supset \operatorname{supp} T,$$

and so

$$|\langle T_t, K(x-t, y)\rangle| \leqslant c\|K(x-t, y)\|_K^m \leqslant c_1 y^{-s}, \qquad x \in K.$$

THEOREM 24.2. *For each $T \in \mathfrak{D}'(\mathbf{R}^n)$, there exists an $f \in H$ which extends continuously by 0 to $\mathbf{R}_+^{n+1} \cup \{\mathbf{R}^n \setminus \operatorname{supp} T\}$ and which is a harmonic representation of T, i.e.*

$$\lim_{\varepsilon \to +0} \int_{y=\varepsilon} f(x, y)\varphi(x)\,dx = \langle T, \varphi\rangle, \qquad \varphi \in \mathfrak{D}(\mathbf{R}^n). \tag{24.1}$$

If $f \in H$ represents 0, then $f \in K$. Every $f \in H$ represents some $T_f \in \mathfrak{D}'$ via (24.1). Thus $\mathfrak{D}' \cong H/K$.

THEOREM 24.2 generalizes Theorem 14.2 to the case of an arbitrary $\mathbf{R}^n$ and an arbitrary $T \in \mathfrak{D}'$. Thus, for distributions defined on hyperplanes, it is more convenient to construct their representations by means of the Poisson kernel rather than the Martinelli-Bochner kernel. The latter can be used for distributions defined on smooth hypersurfaces in $\mathbf{C}^n$.

The first part of Theorem 24.2 has in fact been proved already (Theorem 14.2). We proceed to the proof of the second part.

LEMMA 24.3. *Let $f \in H$. Then, for any bounded domain $\Omega \subset \mathbf{R}^n$, there exist an index p and an $F \in H$ which can be extended continuously to $\mathbf{R}_+^{n+1} \cup \Omega$, such that $f = \partial^{2p}F/\partial y^{2p}$.*

For $n = 1$, a similar lemma for holomorphic functions is due to Tillmann [1].

PROOF. Let Δ be the Laplacian in $\mathbf{R}^{n+1}$, and $\tilde{\Delta}$ that in $\mathbf{R}^n$. Given f harmonic in $\mathbf{R}_+^{n+1}$, consider

$$F_1(x, y) = \int_1^y F(x, \eta)\,d\eta + \varphi(x).$$

Then $\Delta F_1(x, y) = \partial f(x, 1)/\partial y + \tilde{\Delta}\phi(x)$. Choose $\varphi \in C^\infty$ such that $\tilde{\Delta}\varphi = -\partial f(x, 1)/\partial y$ in $\mathbf{R}^n$ (this is possible since the Poisson equation can be solved in $\mathbf{R}^n$). Then $\Delta F_1 = 0$, and $\partial F_1/\partial y = f$. If $f \in H$, then there exist $c > 0$ and $m > 0$ such that

$$|f(x, y)| \leq cy^{-m}, \qquad x \in \overline{\Omega}, 0 < y \leq 1.$$

Hence $F_1 \in H$ and

$$|F_1(x, y)| \leq c_1 y^{-m+1}, \qquad x \in \overline{\Omega}, 0 < y \leq 1.$$

At some step we can find an $F_2 \in H$ which is bounded for $x \in \overline{\Omega}$ and $0 < y \leq 1$ and such that $\partial^s F_2/\partial y^s = f$. The next harmonic primitive $F \in H$ of f will then be continuously extendable to $\mathbf{R}^{n+1} \cup \Omega$, since

$$|F(x, y_1) - F(x, y_2)| \leq c_3|y_1 - y_2|, \qquad x \in \overline{\Omega}, 0 < y_1, y_2 \leq 1.$$

Using this lemma, it is easy to show that the limit in (24.1) exists for any $f \in H$ and $\varphi \in \mathfrak{D}$. Indeed,

$$\begin{aligned}\int_{y=\varepsilon} f(x, y)\varphi(x)dx &= \int_{y=\varepsilon} \frac{\partial^{2p}}{\partial y^{2p}} F(x, y)\varphi(x)\, dx \\ &= (-1)^p \int_{y=\varepsilon} \tilde{\Delta}^p F(x, y)\varphi(x)\, dx \\ &= (-1)^p \int_{y=\varepsilon} F(x, y)\tilde{\Delta}^p\varphi\, dx \xrightarrow[\varepsilon\to+0]{} (-1)^p \int_{\mathbf{R}^n} F(x,0)\tilde{\Delta}^p\varphi(x)\, dx.\end{aligned}$$

Consequently, since $\mathfrak{D}$ is weakly dense in $\mathfrak{D}'$ (Shilov [1], Chapter II, §9.3), the above limit defines a certain distribution $T_f \in \mathfrak{D}$.

Suppose now that, for some $f \in H$, the limit in (24.1) is 0 for all $\varphi \in \mathfrak{D}$. Then, by Lemma 24.3, for each bounded domain $\Omega \subset \mathbf{R}^n$ we have an $F \in H$ and an index p such that $\partial^{2p}F/\partial y^{2p} = f$ and F extends continuously to $\mathbf{R}^{n+1}_+ \cup \Omega$. Since f represents 0, $\tilde{\Delta}^p F(x,0) = 0$ in Ω in the sense of distributions. Hence it follows from the ellipticity of $\tilde{\Delta}^p$ in Ω (see, for example, Schapira [1], Theorem A.22) that $F(x,0)$ is real analytic in Ω. Hence F extends in a C^∞ way to Ω (see Brelot [1], Chapter I, §8). Then

$$f(x, y) = (-1)^p\tilde{\Delta}^p F(x, y) \rightrightarrows (-1)^p\tilde{\Delta}^p F(x,0) = 0,$$

as $y \to +0$.

LEMMA 24.4. *Let f be a polyharmonic function of order m (i.e. $\Delta^m f = 0$) defined in $\mathbf{R}^{n+1}_+$. Then*

$$f(x, y) = f_0(x, y) + yf_1(x, y) + \cdots + y^{m-1}f_{m-1}(x, y), \qquad (24.2)$$

where the f_j, $j = 0, 1, \ldots, m-1$, are harmonic functions in $\mathbf{R}_+^{n+1}$; if f has finite order of growth as $y \to +0$, then the f_j lie in H, and if $f \in C^\infty(\mathbf{R}_+^{n+1} \cup \Omega)$ for an open $\Omega \subset \mathbf{R}^n$, then the f_j are C^∞ in $\mathbf{R}^{n+1} \cup \Omega$.

The proof is by induction on m. If $m = 1$, (24.2) obviously holds. Let $\Delta^m f = 0$, $m > 1$. Then we must have $f = \sum_0^{m-1} y^j f_j$, which would give us

$$\Delta f = \sum_{j=0}^{m-2} y^j \left[(j+2)(j+1) f_{j+2} + 2(j+1) \frac{\partial f_{j+1}}{\partial y} \right].$$

By the induction hypothesis, since Δf is polyharmonic of order $m-1$,

$$\Delta f = \sum_{j=0}^{m-2} y^j \varphi_j,$$

so that we obtain the equations

$$\varphi_j = (j+2)(j+1) f_{j+2} + 2(j+1) \frac{\partial f_{j+1}}{\partial y}, \qquad j = 0, \ldots, m-2,$$

for the f_j; in particular,

$$\varphi_{m-2} = 2(m-1) \frac{\partial f_{m-1}}{\partial y}.$$

As in Lemma 24.3, we seek the solution of this equation in the form of the integral

$$f_{m-1}(x, y) = \frac{1}{2(m-1)} \int_1^y \varphi_{m-2}(x, \eta)\, d\eta + \psi(x).$$

Since

$$\Delta f_{m-1} = \frac{1}{2(m-1)} \frac{\partial \varphi_{m-2}}{\partial y}(x, 1) + \tilde{\Delta} \psi(x) = 0,$$

we find ψ from this equation. Next, from the equation

$$\varphi_{m-3} = (m-1)(m-2) f_{m-1} + 2(m-2) \frac{\partial f_{m-2}}{\partial y}$$

we determine f_{m-2} in a similar way, etc.

It is clear from the proof above that the f_j lie in H if f is of finite order of growth, and that $f_j \in C^\infty(\Omega \cup \mathbf{R}_+^{n+1})$ for all j if $f \in C^\infty(\Omega \cup \mathbf{R}_+^{n+1})$.

COROLLARY 24.4. *If $\Delta^m f = 0$ and f is of finite order of growth as $y \to +0$, then $f(x, y) \to T_f \in \mathfrak{D}'$ in the sense of* (24.1) *as $y \to +0$.*

Thus, polyharmonic functions also represent distributions in $\mathfrak{D}'$, and in this case $T_f = T_{f_0}$.

§25. Multiplication of distributions and its properties

Using the harmonic representations, we can define the product of two distributions by the method of Ivanov [1], [2]. Let H^* be the algebra of functions generated by H, and $K^* \subset H^*$ the subset consisting of functions representing 0 in the sense of (24.1). We shall call the quotient $H^*/K^* = \mathfrak{D}^*$ the space of *hyperdistributions*, and denote by f^* the image of $f \in H^*$ in $\mathfrak{D}^*$, i.e. $f^* = f + K^*$.

The map $A: \mathfrak{D}' \to \mathfrak{D}^*$ defined as $T_f \to f^*$ (f a harmonic representation of T) is an imbedding, since $K \subset K^*$. Hence we shall suppose $\mathfrak{D}' \subset \mathfrak{D}^*$. Let $T, S \in \mathfrak{D}'$, and let f and g be their harmonic representations. Then by the *product* $T \circ S$ of T and S we shall mean the hyperdistribution $(fg)^* = fg + K^* \in \mathfrak{D}^*$.

In the sequel, we shall be interested in the question as to when $T \circ S$ is a distribution, in the sense that the limit (24.1) exists for fg and an arbitrary $\varphi \in \mathfrak{D}$. Note that if this limit exists for fg, then it exists for any function in $fg + K^*$, and coincides with the limit for fg.

We give some properties of this product.

1. *$S \circ T$ is well-defined, i.e. the class $fg + K^*$ is independent of the choice of the representatives f and g.*

PROOF. It suffices to show that, if $f \in K$ and $g \in H$, then $fg \in K^*$. Let $\varphi \in \mathfrak{D}$. It is easy to show that $f = yf_1$ in a neighborhood of the support of φ, where f_1 is a real-analytic function. Hence $f\varphi \in \mathfrak{D}$ for fixed y, and $f\varphi \to 0$ in $\mathfrak{D}$ as $y \to +0$, while g tends to S in $\mathfrak{D}'$ as $y \to +0$; consequently, by a lemma in Shilov [1] (Chapter II, § 9.3),

$$\lim_{\varepsilon \to +0} \int_{\mathbf{R}^n} f(x, \varepsilon)\varphi(x) g(x, \varepsilon)dx = \langle S, 0 \rangle = 0,$$

i.e. $fg \in K^*$.

2. $\operatorname{supp}(T \circ S) \subset \operatorname{supp} T \cap \operatorname{supp} S$. *In particular,* $T \circ S = 0$ *if* $\operatorname{supp} T \cap \operatorname{supp} S = \varnothing$.

Here, the support of a hyperdistribution is defined as follows. We say that $f^* \in \mathfrak{D}^*$ vanishes in an open set Ω in $\mathbf{R}^n$ if, for any $\varphi \in \mathfrak{D}$ with support in Ω,

$$\lim_{\varepsilon \to +0} \int_{y=\varepsilon} f(x, y)\varphi(x)dx = 0, \qquad f \in f^*.$$

Now the support $\operatorname{supp} f^*$ of f^* is by definition the complement of the largest open set in which f^* vanishes. It is easy to verify by means of a partition of unity that this notion is well-defined.

PROOF OF PROPERTY 2. Let $\varphi \in \mathfrak{D}$ with $\operatorname{supp} \varphi \subset \mathbf{R}^n \backslash (\operatorname{supp} T \cap \operatorname{supp} S) = \varnothing$. We may suppose that, for example, $\operatorname{supp} \varphi \cap \operatorname{supp} T = \varnothing$ (by means of a partition of unity it is always possible to pass to functions φ with arbitrarily small support). Then f can be written in the form yf_1 in a neighborhood of supp φ, and the rest of the proof proceeds as for Property 1.

3. *If* $T \in \mathfrak{D}'$ *is defined by a function* $T(x) \in L^p(\text{loc})$, *and* S *by* $S(x)$ *in* $L^q(\text{loc})$, $1/p + 1/q = 1$, *then* $T \circ S = T(x)L(x) \in L^1(\text{loc})$.

PROOF. It can be shown that, corresponding to $S(x)$, there exists $u(x, y) \in H$ such that $u(x, y) \to S(x)$ almost everywhere as $y \to +0$, and for each compact set M

$$\|u(x, y) - S(x)\|_M^q \to 0$$

as $y \to +0$, where

$$\|f\|_M^q = \left(\int_M |f|^q\, dx\right)^{1/q}, \qquad q \geqslant 1.$$

To do this, we must use:

(a) Theorem 1 of Stein [2], (Chapter III, §2.1), asserting that, for every $f \in L^q(\mathbf{R}^n)$, its Poisson integral defines a harmonic function $\hat{f}(x, y)$ such that $\hat{f}(x, y)$ converges to $f(x)$ almost everywhere and $\|\hat{f}(x, y) - f(x)\|_{\mathbf{R}^n}^q \to 0$ as $y \to +0$; and

(b) the procedure in Bremermann [1], §5.9, replacing power-series expansions by expansions in harmonic polynomials.

Since $S(x) \in L^q(\text{loc})$ and $T(x) \in L^p(\text{loc})$, we will have $S(x)T(x) \in L^1(\text{loc})$. Let $u(x, y)$, $v(x, y)$ and $w(x, y)$ be the respective harmonic extensions to $\mathbf{R}_+^{n+1}$ with the properties mentioned above. Then u, v and w are harmonic representations of $T(x)$, $S(x)$ and $T(x)S(x)$. Furthermore $uv - w \in K^*$. Indeed, let $\varphi \in \mathfrak{D}$ with supp $\varphi \subset M$. Then

$$\left|\int_{\mathbf{R}^n} (uv - w)\varphi\, dx\right| \leqslant c\int_M |uv - w|dx \leqslant c\left(\|uv - ST\|_M^1 + \|w - ST\|_M^1\right),$$

but $\|uv - ST\|_M^1 \to 0$ as $y \to +0$, since $\|u - T\|_M^p$ and $\|v - S\|_M^q$ tend to 0.

4. *The derivatives of hyperdistributions are defined as follows*:

$$\frac{\partial}{\partial x_k} f^* = \left(\frac{\partial}{\partial x_k} f\right)^* = \frac{\partial f}{\partial x_k} + K^*.$$

This definition makes sense, since $\partial f/\partial x_k \in K^*$ for $f \in K^*$. *For the derivatives of the product of two distributions, we have the Leibniz rule*

$$\frac{\partial}{\partial x_k}(T \circ S) = \frac{\partial T}{\partial x_k} \circ S + T \circ \frac{\partial S}{\partial x_k}.$$

This follows immediately from the fact that if f represents T, then $\partial f/\partial x_k$ represents $\partial T/\partial x_k$.

5. *The product of distributions is compatible with multiplication by multipliers, i.e. if $\alpha \in C^\infty(\mathbf{R}^n)$ and $T \in \mathfrak{D}'$, then $\alpha \circ T = \alpha T$.*

PROOF. Since $\alpha \in C^\infty$, its harmonic representation $g(x, y) \in H$ extends continuously to $\mathbf{R}^{n+1}_+ \cup \mathbf{R}^n$ (solution of the Dirichlet problem for the half-space). Hence $g \in C^\infty(\mathbf{R}^{n+1}_+ \cup \mathbf{R}^n)$ (Brelot [1], Chapter I, §8), so that $\varphi(x)g(x, y) \to \varphi(x)\alpha(x)$ in $\mathfrak{D}$ as $y \to +0$, for any $\varphi \in \mathfrak{D}$. On the other hand, $f(x, y) \to T$ in $\mathfrak{D}'$ as $y \to +0$, so that, by the lemma already mentioned (Shilov [1], Chapter II, §9.3),

$$\int_{y=\varepsilon} f(x, y)\varphi(x)g(x, y)\,dx \underset{\varepsilon\to+0}{\longrightarrow} \langle T, \alpha\varphi\rangle = \langle \alpha T, \varphi\rangle.$$

6. *If $f \in H$ is a homogeneous function of degree p, then its boundary value as $y \to +0$, i.e. T_f, is also homogeneous of degree p.*

Indeed,

$$\langle T_f, \varphi(tx)\rangle = \lim_{\varepsilon\to+0}\int_{y=\varepsilon} f(x, y)\varphi(tx)\,dx = \lim_{\varepsilon\to+0}\frac{1}{t^n}\int_{y=\varepsilon t} f\left(\frac{x}{t}, \frac{y}{t}\right)\varphi(x)\,dx$$
$$= \frac{1}{t^{p+n}}\lim_{\varepsilon\to+0}\int_{y=\varepsilon t} f(x, y)\varphi(x)\,dx = \frac{1}{t^{p+n}}\langle T_f, \varphi\rangle.$$

If $T \in \mathcal{O}'_{-n}$ is homogeneous of degree p, then it is not hard to see that $T^*(x, y)$ is a homogeneous function of degree p, since

$$T^*(tx, ty) = \langle T_u, K(u - tx, ty)\rangle = \frac{1}{t^n}\left\langle Tu, K\left(\frac{u}{t} - x, y\right)\right\rangle$$
$$= t^p\langle T_u, K(u - x, y)\rangle = t^p T^*(x, y).$$

By means of the general procedure for constructing a harmonic representation, it can be shown that, if $T \in \mathfrak{D}'$ is homogeneous of degree p, then T admits a harmonic representation which is homogeneous of degree p.

It follows that, if $T, S \in \mathfrak{D}'$ are homogeneous of degrees p and q respectively, and $T \circ S \in \mathfrak{D}'$, then $T \circ S$ is homogeneous of degree $p + q$.

7. *Let $x' = (x_1, \ldots, x_m)$, $x'' = (x_{m+1}, \ldots, x_n)$, and $T \in \mathfrak{D}'(\mathbf{R}^m)$, $S \in \mathfrak{D}'(\mathbf{R}^{n-m})$. Let $f(x', y)$ and $g(x'', y)$ be harmonic representations of T and S in $\mathbf{R}^{m+1}_+$ and $\mathbf{R}^{n-m+1}_+$ respectively. Then*

$$(f + K_{x'})\cdot(g + K_{x''}) \subset h + K^*$$

where h is a harmonic representation of the tensor product $T \times S$. Thus multiplication of distributions is compatible with the formation of products.

PROOF. It suffices to show that the boundary value of $f(x', y)g(x'', y)$ in the sense of (24.1) is $T \times S$. Consider the integral

$$\int_{y=\varepsilon} g(x'', y)\varphi(x)dx'' = \psi(x', \varepsilon), \qquad \varphi \in \mathfrak{D}.$$

We have $\psi(x', \varepsilon) \in \mathfrak{D}(\mathbf{R}^m)$ and $\psi(x', \varepsilon) \to \langle S, \varphi(x)\rangle = \psi(x')$ in $\mathfrak{D}(\mathbf{R}^m)$ as $\varepsilon \to +0$. Since $\psi(x', \varepsilon) \rightrightarrows \psi(x')$ as $\varepsilon \to +0$ (Lemma 24.3), and

$$D^\alpha_{x'}\psi(x', \varepsilon) = \int_{y=\varepsilon} g(x'', y)D^\alpha_{x'}\varphi(x)dx'',$$

we have, by the same lemma, $D^\alpha_{x'}\psi(x, \varepsilon) \rightrightarrows D^\alpha_{x'}(\psi(x'))$ as $\varepsilon \to +0$. Now we apply the lemma from Shilov [1] (Chapter II, §9.3) once again and conclude

$$\begin{aligned}\int_{y=\varepsilon} f(x', y)g(x'', y)\varphi(x)dx &= \int_{y=\varepsilon} f(x', y)\psi(x', y)dx' \underset{\varepsilon \to +0}{\longrightarrow} \langle T_{x'}, \psi(x')\rangle \\ &= \langle T_{x'}, \langle S_{x''}, \varphi(x)\rangle\rangle = \langle T \times S, \varphi(x)\rangle.\end{aligned}$$

Thus the definition of product which we have given has several natural properties; in particular, we will not have counterexamples of the Itano kind with our definition.

§26. Examples of products of distributions

Consider the vector space $\mathfrak{M}$ of functions, generated by functions of the form

$$f(x, y) = \frac{P_k(x, y)}{\left(|x|^2 + y^2\right)^{k-m+(n+1)/2}}, \qquad k \geqslant 0, m \geqslant 1, \tag{26.1}$$

where $P_k(x, y)$ is a homogeneous polynomial of degree k and $\Delta^m P_k = 0$. These functions are polyharmonic of order m; actually they are the Kelvin transforms of the $P_k(x, y)$. We first elucidate what kind of distributions they represent.

LEMMA 26.1. *Suppose f is a function of the form* (26.1) *lying in $\mathfrak{M}$, and $k + 1 - 2m > 0$. Then*

$$\int_{\mathbf{R}^n} \frac{P_k(x, y)x^\alpha dx}{\left(|x|^2 + y^2\right)^{k-m+(n+1)/2}} = 0$$

for all monomials x^α with $\|\alpha\| < k + 1 - 2m$.

PROOF. Let

$$I(y) = \int_{|x| \leqslant 1} \frac{P_k(x, y)x^\alpha dx}{\left(|x|^2 + y^2\right)^{k-m+(n+1)/2}}, \qquad \|\alpha\| < k + 1 - 2m.$$

We shall show that the limit

$$\lim_{y\to+0} I(y) = c$$

exists and is finite. Indeed, by Lemma 24.4, we can write f in the form

$$f = f_0 + yf_1 + \cdots + y^{m-1}f_{m-1};$$

then

$$\lim_{y\to+0} I(y) = \sum_{j=0}^{m-1} \lim_{y\to+0} y^j \int_{|x|\leqslant 1} f_j(x, y)x^\alpha\, dx.$$

According to Lemma 24.4, the f_j are C^∞ on $\mathbf{R}_+^{n+1} \cup (\mathbf{R}^n\setminus\{0\})$. We now show that the limit

$$\lim_{y\to+0} \int_{|x|\leqslant 1} f_j(x, y)x^\alpha dx$$

exists.

By Lemma 24.3, there exists $F_j(x, y) \in H$ extending continuously to $\mathbf{R}_+^{n+1} \cup \{x \in \mathbf{R}^n\colon |x| < 2\}$ and such that

$$\frac{\partial^{2p}}{\partial y^{2p}} F_j(x, y) = f_j(x, y).$$

Then $f_j(x, y) = (-1)^p\tilde{\Delta}^p F_j(x, y)$, and

$$\int_{|x|\leqslant 1} \frac{\partial}{\partial x_k} F_j(x, y)\varphi(x)dx$$

$$= -\int_{|x|\leqslant 1} F_j(x, y)\frac{\partial \varphi}{\partial x_k} dx + (-1)^{k-1}\int_{|x|=1} F_j\varphi dx \qquad [1].$$

Thus

$$\int_{|x|\leqslant 1} f_j(x, y)x^\alpha dx = (-1)^p \int_{|x|\leqslant 1} F_j(x, y)\tilde{\Delta}^p x^\alpha dx + \int_{|x|=1} DF_j \cdot x^\alpha d\sigma,$$

where D is a differential operator, and $d\sigma$ the surface element of the sphere. We need only observe now that the F_j, like the f_j, are C^∞ on $\mathbf{R}_+^{n+1} \cup (\{|x| < 2\}\setminus\{0\})$ (see the proof of Lemma 24.3).

Thus the limit

$$\lim_{y\to+0} I(y) = c$$

exists. We now transform the variables in $I(y)$ by setting $x = r\xi$, $|\xi| = 1$. Then

$$I(y) = \sum_{s\geqslant 0} y^s \int_0^1 \frac{r^{\|\alpha\|+n-1+k-s}}{(r^2+y^2)^{k-m+(n+1)/2}} dr \int_{|\xi|=1} P_k^s(\xi)\xi^\alpha d\sigma$$

where $P_k(x, y) = \Sigma_{s\geq 0} y^s P_k^s(x)$. If we further set $r = yt$, then

$$I(y) = y^{\|\alpha\|-k-1+2m} \int_0^{1/y} \sum_{s\geq 0} \frac{t^{\|\alpha\|+n-1+k-s}dt}{(1+t^2)^{k-m+(n+1)/2}} \int_{|\xi|=1} P_k^s(\xi)\xi^\alpha d\sigma. \quad (26.2)$$

Since $I(y)$ has a limit as $y \to +0$, we necessarily obtain from (26.2) that

$$J = \lim_{y\to+0} I(y) = \int_0^\infty \sum_{s\geq 0} \frac{t^{\|\alpha\|+n-1+k-s}}{(1+t^2)^{k-m+(n+1)/2}} dt \int_{|\xi|=1} P_k^s(\xi)\xi^\alpha d\sigma = 0.$$

The improper integral above is convergent, since $\|\alpha\| < k + 1 - 2m$.

It is easy to see that

$$\int_{\mathbf{R}^n} \frac{P_k(x, y)x^\alpha dx}{(|x|^2 + y^2)^{k-m+(n+1)/2}} = y^{\|\alpha\|-k-1-2m} J = 0.$$

LEMMA 26.2. *Suppose an f of the form* (26.1) *lies in* $\mathfrak{M}$ *and* $k + 1 - 2m \geq 0$. *Then*

$$I = \text{p.v.}\int_{\mathbf{R}^n} \frac{P_k(x, y)x^\alpha dx}{(|x|^2 + y^2)^{k-m+(n+1)/2}} = c_\alpha, \qquad \|\alpha\| = k + 1 - 2m,$$

where

$$c_\alpha = \sum_{s\geq 1} \frac{\pi^{n/2}\Gamma(s - 1/2)\tilde{\Delta}^{k+1-s-m}(P_k^{2s-1}(x)x^\alpha)}{2^{2(k+1-s-m)}(k + 1 - s - m)!\Gamma(k - m + (n + 1)/2)}. \quad (26.3)$$

PROOF. Consider the integral

$$I(y) = \int_{|x|\leq 1} \frac{P_k(x, y)x^\alpha dx}{(|x|^2 + y^2)^{k-m+(n+1)/2}}.$$

Transforming to polar coordinates as in Lemma 26.1, we get

$$I(y) = \sum_{s\geq 0} \int_0^{1/y} \frac{t^{2k+n-s-2m}dt}{(1+t^2)^{k-m+(n+1)/2}} \int_{|\xi|=1} P_k^s(\xi)\xi^\alpha d\sigma.$$

The polynmials P_k^s have degrees $k - s$; hence the $P_k^s\xi^\alpha$ have degrees $2k + 1 - 2m - s$. Thus, if s is even, the degree of the homogeneous polynomial $P_k^s\xi^\alpha$ is odd, and so

$$\int_{|\xi|=1} P_k^s\xi^\alpha d\sigma = 0.$$

Then,

$$I(y) = \sum_{s\geq 1} \int_0^{1/y} \frac{t^{2k+n+1-2s-2m}dt}{(1+t^2)^{k-m+(n+1)/2}} \int_{|\xi|=1} P_k^{2s-1}(\xi)\xi^\alpha d\sigma.$$

Clearly $I(y) \to I$ as $y \to +0$, so that

$$J = \sum_{s \geqslant 1} \int_0^\infty \frac{t^{2k+n+1-2s-2m} dt}{(1+t^2)^{k-m+(n+1)/2}} \int_{|\xi|-1} P_k^{2s-1}(\xi)\xi^\alpha d\sigma.$$

But

$$\int_0^\infty \frac{t^{2k+n+1-2s-2m} dt}{(1+t^2)^{k-m+(n+1)/2}} = \frac{1}{2} B\left(k - s - m + 1 + \frac{n}{2}, s - \frac{1}{2}\right),$$

and

$$\int_{|\xi|=1} P_k^{2s-1} \xi^\alpha d\sigma = \frac{2\pi^{n/2} \tilde{\Delta}^{k+1-s-m}\left(P_k^{2s-1}(\xi)\xi^\alpha\right)}{2^{2(k+1-s-m)}(k+1-s-m)!\Gamma(k+1-s-m+n/2)}.$$

(This follows from the Gauss representation of homogeneous polynomials; see Sobolev [1], Chapter 11). Hence $I = c_\alpha$.

Observe that if P_k expands only in even powers of y, then $c_\alpha = 0$ for all α such that $\|\alpha\| \leqslant k + 1 - 2m$.

Let

$$h(x) = \frac{Q_k(x)}{|x|^{2k-2m+n+1}},$$

where Q_k is an arbitrary homogeneous polynomial of degree k in $\mathbf{R}^n$. We denote by h the distribution obtained by regularizing $h(x)$, i.e.

$$\begin{aligned} \langle \hat{h}, \varphi \rangle &= \text{f.p.} \int_{\mathbf{R}^n} \frac{Q_k(x)\varphi(x)dx}{|x|^{2k-2m+n+1}} \\ &= \text{p.v.} \int_{\mathbf{R}^n} \frac{Q_k(x)\left(\varphi(x) - \sum_{\|\alpha\| \leqslant k+1-2m} (1/\alpha!)(\partial^\alpha \varphi/\partial x^\alpha)(0)x^\alpha\right)dx}{|x|^{2k-2m+n+1}} \end{aligned}$$

if $k + 1 - 2m \geqslant 0$, where the abbreviations "f.p." and "p.v." signify "(Hadamard) finite part" and "(Cauchy) principal value".

If $k + 1 - 2m < 0$, then

$$\langle \hat{h}, \varphi \rangle = \int_{\mathbf{R}^n} h(x)\varphi(x)dx, \qquad \varphi \in \mathfrak{D}.$$

In other words, $\hat{h}$ is the Hadamard finite part of the integral

$$\int_{\mathbf{R}^n} h(x)\varphi(x)dx.$$

LEMMA 26.3. *Let Q be any polynomial in $\mathbf{R}^n$. Then*

$$\tilde{Q}(x, y) = \sum_{l \geqslant 0} \frac{(-1)^l y^{2l} \tilde{\Delta}^l Q(x)}{(2l)!}$$

is a harmonic extension of Q to all of $\mathbf{R}^{n+1}$.

This follows from direct computation.

If

$$h(x) = \frac{\tilde{Q}_k(x, y)}{\left(|x|^2 + y^2\right)^{k-m+(n+1)/2}},$$

then we denote by $\tilde{h}(x, y)$ the function of the form

$$\tilde{h}(x, y) = \frac{\tilde{Q}_k(x, y)}{\left(|x|^2 + y^2\right)^{k-m+(n+1)/2}},$$

in $\mathfrak{M}$.

THEOREM 26.4. *The distributions $\hat{h}$ and $T_{\tilde{h}}$ are identical.*

This follows immediately from Lemmas 26.1–26.3 and the definitions of $\hat{h}$ and $T_{\tilde{h}}$.

Let $\mathfrak{M}_1$ (respectively $\mathfrak{M}_2$) denote the subset of $\mathfrak{M}$ consisting of functions of the form (26.1) with P_k expanding only in even (odd) powers of y. It is easy to verify that $\operatorname{span}(\mathfrak{M}_1 + \mathfrak{M}_2) = \mathfrak{M}$.

Theorem 26.4 implies that the boundary values of functions f in $\mathfrak{M}_1$ in the sense of (24.1) are the distributions $(f(x,0))\hat{}$.

THEOREM 26.5. *If $f \in \mathfrak{M}_2$, then*

$$T_f = \begin{cases} (-1)^{k+1+2m} \displaystyle\sum_{\|\alpha\| = k+1-2m} \frac{c_\alpha}{\alpha!} \delta^{(\alpha)}, & \text{if } k+1-m \geqslant 0, \\ 0, & \text{if } k+1-m < 0, \end{cases}$$

where the c_α are defined by (26.3) *and $\delta^{(\alpha)} = \partial^\alpha \delta / \partial x^\alpha$, δ being the Dirac δ-function.*

This is immediate from Lemmas 26.1 and 26.2. We have thus a complete description of the class of distributions representable by the functions in $\mathfrak{M}$.

EXAMPLE 1. Let $f(x, y) = y/(|x|^2 + y^2)^{(n+1)/2}$. Then

$$T_f = \pi^{(n+1)/2} \delta / \Gamma((n+1)/2).$$

We now consider the question: when is the product of two functions in $\mathfrak{M}$ again im $\mathfrak{M}$? Let

$$f = \frac{P_k(x, y)}{\left(|x|^2 + y^2\right)^{k-m+(n+1)/2}}, \qquad g = \frac{Q_l(x, y)}{\left(|x|^2 + y^2\right)^{l-s+(m+1)/2}},$$

then

$$fg = \frac{P_k Q_l}{\left(|x|^2 + y^2\right)^{l+k+(n+1)/2-(m+s-(n+1)/2)}}.$$

Hence we see that $fg \in \mathfrak{M}$ if and only if n is odd $(n + 2p - 1)$, $m + s \geqslant p$, and $\Delta^{m+s-p}(P_k Q_l) = 0$. Thus the class of such functions f and g is fairly large.

THEOREM 26.6. *Suppose f, g and fg lie in $\mathfrak{M}$. Then*:
a) *if $f, g \in \mathfrak{M}_1$, then $fg \in \mathfrak{M}_1$ and* $(f(x,0))\hat{}\circ(g(x,0))\hat{} = (fg(x,0))\hat{}$;
b) *if $f, g \in \mathfrak{M}_2$, then $T_f \circ T_g = 0$;*
c) *if $f \in \mathfrak{M}_1$ and $g \in \mathfrak{M}_2$, then $fg \in \mathfrak{M}_2$ and*

$$(-1)^{k-2p-2m} \hat{f}(x,0) \circ \sum_{\|\alpha\| = l+1-2s} \frac{c_\alpha}{\alpha!} \delta^{(\alpha)} = \sum_{\|\beta\| = l+k+2p-2m-2s-1} \frac{c_\beta}{\beta!} \delta^{(\beta)}.$$

For $n = 1$, identities of this type have been obtained by Ivanov [3].

EXAMPLE 2. Let $f = (p-1)!y/\pi^p(|x|^2 + y^2)^p$. Then $T_f = \delta$. Let $h(x) = Q_k(x)/|x|^{2k}$ with $\tilde{\Delta} Q_k = 0$; then

$$\hat{h} \circ \delta = (-1)^k \sum_{\|\alpha\| = k} \frac{(p-1)!\tilde{\Delta}^k(Q_k x^\alpha)}{k!\,(k+p-1)!} \delta^{(\alpha)}.$$

If $h = x_j/|x|^2$, then

$$\left(\frac{x_j}{|x|^2}\right)^{\wedge} \circ \delta = -\frac{1}{2p} \frac{\partial}{\partial x_j} \delta.$$

For $p = n = 1$, this identity reduces to the familiar identity $P(1/x) \cdot \delta = -\delta_2'$.

SUPPLEMENT TO THE BRIEF HISTORICAL SURVEY

§16 contains the solution to Problem 3 (see "Brief historical survey and open problems" which follows Chapter IV). All the results of this section are due to Kytmanov [2]–[4]. A special case of Problem 3 (viz., Problem 2) is discussed in §17. Theorem 17.1 was proved for the ball in $\mathbf{C}^n$ by Aronov [2], and in the general case by Aronov and Kytmanov [1]. Theorem 17.2 and Corollaries 17.3 and 17.4 were obtained by Aĭzenberg and Kytmanov [1], [2]. Corollary 17.7 was obtained for the ball by Agranovskiĭ and Val'skiĭ [1], and for arbitrary domains by Stout [1]. Theorems 17.9 and 17.10 were proved by Kytmanov [1]. The results of §18 are due to Aĭzenberg and Dautov [1]. All the assertions of §19 except for Theorems 19.4 and 19.5 are due to Kytmanov [8]. Theorems 19.4 and 19.5 were proved by Romanov [1]; Theorem 19.5 is in fact true for arbitrary domains with smooth boundary in $\mathbf{C}^n$ (Romanov [2]). In §20 differential criteria for the existence of holomorphic extensions of functions are presented, which are criteria different from the tangential Cauchy-Riemann equations. Theorem 20.1 was proved for domains D in Kähler manifolds with $\partial D \in C^\infty$ and $f \in C^\infty(\overline{D})$ by Folland and Kohn [1]; in the form stated here, it is contained in Aronov and Kytmanov [1]. Theorem 20.4 is a reformulation of a result of Forelli [1]. The remaining results of §20 are due to Kytmanov [7].

The strengthening of Theorem 6.1 (Theorem 21.5) was obtained by Dautov (see Aĭzenberg and Yuzhakov [1], Theorem 26.1). The results of §22 are due to Dautov. Weighted formulas for solving the $\bar{\partial}$-problem and weighted estimates (Theorem 23.1) were proved by Dautov and Khenkin [1], [2]. Theorem 21.6 is due to Khenkin [5]. For $\alpha = 0$, Theorem 23.2 (for functions of the Nevanlinna class) was proved by Khenkin [4]–[6] and Skoda [1], [2]; for $\alpha > 0$ (functions of the Nevanlinna-Dzhrbashy class) it was proved by Dautov and Khenkin [1], [2]. In these papers, there is also a generalization of Theorem 23.2 to strictly pseudoconvex domains in manifolds.

Chapter VII is devoted to the solution of Problem 13. The results of this chapter are due to Kytmanov [5], [6]. We note that, for $n = 1$, Theorem 24.1 is proved in Bremermann [1] (p. 49). That part of Theorem 24.2 which proves that the boundary value of a harmonic function of class H is a distribution can be deduced from a theorem of Roĭtberg [1] which states that, if L is an elliptic operator of order $2m$ in a bounded domain $\Omega \subset \mathbf{R}^n$ with sufficiently smooth boundary, then every solution f of the equation $Lf = 0$ with finite-order singularity on $\partial\Omega$ defines a certain distribution on $\partial\Omega$. Theorem 24.2 gives a method for constructing this distribution. A different method of multiplying higher-dimensional distribution by means of analytic representations has been proposed by Ivanov and Verzhbalovich [1]. We note that, with this method as with ours, it would not be possible to have counterexamples like those of Itano [1].

BIBLIOGRAPHY

(Items marked * were added for the English translation.)

M. L. Agranovskiĭ and R. E. Val'skiĭ

1*. *Maximality of invariant algebras of functions*, Sibirsk. Mat. Zh. **12** (1971), 3–12; English transl. in Siberian Math. J. **12** (1971).

L. A. Aĭzenberg

1. *Linear convexity in $\mathbf{C}^n$ and the separation of the singularities of holomorphic functions*, Bull. Acad. Polon. Sci. Sér. Sci. Math. Astronom. Phys. **15** (1967), 487–495. (Russian)
2. *The expansion of holomorphic functions of several complex variables in partial fractions*, Sibirsk. Mat. Zh. **8** (1967), 1124–1142; English transl. in Siberian Math. J. **8** (1967).
3. *On Temlyakov integrals and the boundary properties of analytic functions of two complex variables*, Dokl. Akad. Nauk SSSR **120** (1958), 935–938. (Russian)
4. *On Temlyakov integrals and the boundary properties of analytic functions of several complex variables*, Moskov. Oblast. Ped. Inst. Uchen. Zap. **77** (Trudy Kafedr Mat. 5) (1959), 13–35. (Russian)
5. *Integral representations of holomorphic functions of several complex variables*, Trudy Moskov. Mat. Obshch. **21** (1970), 3–26; English transl. in Trans. Moscow Math. Soc. **21** (1970).
6. *On differential forms that are orthogonal to holomorphic functions under integration over the boundary of a domain $D \subset \mathbf{C}^n$*, All-Union Conf. Theory of Functions of a Complex Variable (Entire and Meromorphic Functions and Functions of Several Variables), Abstracts of Reports, Fiz.-Tekhn. Inst. Nizkikh Temperatur Akad. Nauk Ukrain. SSR, Kharkov, 1971, pp. 10–12. (Russian)
7. *On $\bar{\partial}$-closed exterior differential forms of type $(p, n-1)$ in the space $\mathbf{C}^n$*, Dokl. Akad. Nauk SSSR **209** (1973), 1009–1012; English transl. in Soviet Math. Dokl. **14** (1973).
8. *Polynomials orthogonal to holomorphic functions of several complex variables and an analogue of the Riesz theorem*, Dokl. Akad. Nauk SSSR **199** (1971), 255–257; English transl. in Soviet Math. Dokl. **12** (1971).
9. *Polynomials that are orthogonal to holomorphic functions of several complex variables, and a generalization of the theorems of Hartogs and F. and M. Riesz*, Holomorphic Functions of Several Complex Variables (Proc. Sem., 1972), Inst. Fiz. Sibirsk. Otdel. Akad. Nauk SSSR, Krasnoyarsk, 1972, pp. 5–19. (Russian)
10. *Integral representations of holomorphic functions of several complex variables*, Dokl. Akad. Nauk SSSR **155** (1964), 9–12; English transl. in Soviet Math. Dokl. **5** (1964).

L. A. Aĭzenberg and Sh. A. Dautov

1*. *Holomorphic functions of several complex variables with nonnegative real part. Traces of holomorphic and pluriharmonic functions on the Shilov boundary*, Mat. Sb. **99(141)** (1976), 342–355; English transl. in Math. USSR Sb. **28** (1976).

L. A. Aĭzenberg and A. P. Yuzhakov

1*. *Integral representations and residues in multidimensional complex analysis*, "Nauka", Novosibirsk, 1979. (Russian)

L. A. Aĭzenberg and A. M. Kytmanov

1. *Representation and multiplication of distributions from* $\mathbf{R}^{2n-1}$ *with the aid of* $\bar{\partial}$*-closed exterior differential forms of type* $(n, n-1)$, Manuscript No. 2690-74, deposited at VINITI, 1974; abstract published in Sibirsk. Mat. Zh. **16** (1975), 409; English transl. in Siberian Math. J. **16** (1975).

A. Andreotti

1. *E. E. Levi convexity and H. Lewy problem*, Proc. Internat. Congr. Math. (Nice, 1970), Vol. 2, Gauthier-Villars, Paris, 1971, pp. 607–611.

Aldo Andreotti and C. Denson Hill

1. *E. E. Levi convexity and the Hans Lewy problem*. I, Ann. Scuola Norm. Sup. Pisa (3) **26** (1972), 325–363.
2. *E. E. Levi convexity and the Hans Lewy problem*. II, Ann. Scuola Norm. Sup. Pisa (3) **26** (1972), 747–806.

V. I. Arnol'd

1. *Ordinary differential equations*, "Nauka", Moscow, 1971; English transl., M.I.T. Press, Cambridge, Mass., 1973.

A. M. Aronov

1. *The Martinelli-Bochner integral formula for exterior differential forms*, Holomorphic Functions of Several Complex Variables (Proc. Sem., 1972), Inst. Fiz. Sibirsk. Otdel. Akad. Nauk SSSR, Krasnoyarsk, 1972, pp. 167–170. (Russian)
2. *Functions that can be represented by a Martinelli-Bochner integral*, Some Properties of Holomorphic Functions of Several Complex Variables, Inst. Fiz. Sibirsk. Otdel. Akad. Nauk SSSR, Krasnoyarsk, 1973, pp. 35–39. (Russian)

3*. *On the boundary values of the derivatives of an integral of Bochner-Martinelli type*, On Holomorphic Functions of Several Complex Variables, Inst. Fiz. Sibirsk. Otdel. Akad. Nauk SSSR, Krasnoyarsk, 1976, pp. 20–28. (Russian)

A. M. Aronov and Sh. A. Dautov

1. *Integral representations for complex differential forms*, Sibirsk. Mat. Zh. **13** (1972), 739–747; English transl. in Siberian Math. J. **13** (1972).
2. *On a result of B. Weinstock*, Some Properties of Holomorphic Functions of Several Complex Variables, Inst. Fiz. Sibirsk. Otdel. Akad. Nauk SSSR, Krasnoyarsk, 1973, pp. 193–195. (Russian)

A. M. Aronov and A. M. Kytmanov

1. *The holomorphy of functions that are representable by the Martinelli-Bochner integral*, Funktsional. Anal. i Prilozhen. **9** (1975), no. 3, 83–84; English transl. in Functional Anal. Appl. **9** (1975).

Stefan Banach

1. *Théorie des opérations linéaires*, PWN, Warsaw, 1932; reprint, Chelsea, New York, 1955.

M. Brelot

1*. *Éléments de la théorie classique du potential*, 2nd ed., Centre de Documentation, Paris, 1961.

Hans Bremermann

1. *Distributions, complex variables and Fourier transforms*, Addison-Wesley, 1965.

E. M. Chirka

1. *Approximations by holomorphic functions on smooth manifolds in* $\mathbf{C}^n$, Mat. Sb. **78(120)** (1969), 101–123; English transl. in Math. USSR Sb. **7** (1969).

Sh. A. Dautov

1. *On forms orthogonal to holomorphic functions with respect to integration over the boundary of strictly pseudoconvex domains*, Dokl. Akad. Nauk SSSR **203** (1972), 16–18; English transl. in Soviet Math. Dokl. **13** (1972).
2. *On forms that are orthogonal to holomorphic functions and related problems*, Funktsional. Anal. i Prilozhen. **6** (1972), no. 4, 81–82; English transl. in Functional Anal. Appl. **6** (1972).
3. *$\bar\partial$-closed forms of type $(p, n-1)$ as an analogue of holomorphic functions of a single complex variable*, Holomorphic Functions of Several Complex Variables (Proc. Sem., 1972), Inst. Fiz. Sibirsk. Otdel. Akad. Nauk SSSR, Krasnoyarsk, 1972, pp. 21–36. (Russian)
4. *The general form of the integral representation of holomorphic functions in* $\mathbf{C}^n$, Holomorphic Functions of Several Complex Variables (Proc. Sem., 1972), Inst. Fiz. Sibirsk. Otdel. Akad. Nauk SSSR, Krasnoyarsk, 1972, pp. 37–45. (Russian)

Sh. A. Dautov and G. M. Kenkin

1*. *Zeros of holomorphic functions of finite order and weighted estimates for solutions of the $\bar\partial$-equation*, Preprint No. 8M, Inst. Phys. Siberian Division Acad. Sci. USSR, Krasnoyarsk, 1977. (Russian) R Zh Mat. **1978** #7Б257.

2*. *Zeros of holomorphic functions of finite order and weighted estimates for solutions of the $\bar\partial$-equation*, Mat. Sb. **107(149)** (1978), 163–174; English transl. in Math. USSR Sb. **35** (1979), no. 4.

Sh. A. Dautov and A. M. Kytmanov

1. *The boundary values of an integral of Bochner-Martinelli type*, Some Properties of Holomorphic Functions of Several Complex Variables, Inst. Fiz. Sibirsk. Otdel. Akad. Nauk SSSR, Krasnoyarsk, 1973, pp. 49–54. (Russian)

Nelson Dunford and Jacob T. Schwartz

1. *Linear operators*. Vol. I, Interscience, 1958.

Gaetano Fichera

1. *Caratterizzazione della traccia, sulla frontiera di un campo, di una funzione analitica di più variabili complesse*, Atti Accad. Naz. Lincei Rend. Cl. Sci. Fis. Mat. Nat. (8) **22** (1957), 706–715.

G. M. Fikhtengol'ts

1. *A course in differential and integral calculus*. Vol. I, 4th ed., Fizmatgiz, Moscow, 1958; German transl. of 5th ed., VEB Deutscher Verlag Wiss., Berlin, 1968.

Wolfgang Fischer and Ingo Lieb

1*. *Lokale Kerne und beschränkte Lösungen für den $\bar\partial$-operator auf q-convexen Gebieten*, Math. Ann. **208** (1974), 249–265.

G. B. Folland and J. J. Kohn

1*. *The Neumann problem for the Cauchy-Riemann complex*, Ann of Math. Studies, No. 75, Princeton Univ. Press, Princeton, N.J., and Univ. of Tokyo Press, Tokyo, 1972.

Frank Forelli

1*. *Pluriharmonicity in terms of harmonic slices*, Math. Scand. **41** (1977), 358–364.

B. A. Fuks

1. *Introduction to the theory of analytic functions of several complex variables*, Fizmatgiz, Moscow, 1962; English transl., Transl. Math. Mono., vol. 8, Amer. Math. Soc., Providence, R.I., 1963.
2. *Special chapters in the theory of analytic functions of several complex variables*, Fizmatgiz, Moscow, 1963; English transl., Transl Math. Mono., vol. 14, Amer. Math. Soc., Providence, R.I., 1965.

Theodore W. Gamelin

1. *Uniform algebras*, Prentice-Hall, Englewood Cliffs, N.J., 1969.

I. M. Gel'fand and G. E. Shilov

1. *Generalized functions.* Vol. II: *Spaces of test functions*, Fizmatgiz, Moscow, 1958; English transl., Academic Press, 1968, and Gordon and Breach, 1968.

N. M. Gunther [Gyunter]

1. *La théorie du potential et ses applications aux problèmes fondamentaux de la physique mathématique*, Gauthier-Villars, Paris, 1934; Russian transl., GITTL, Moscow, 1953; English transl. (of the Russian), Ungar, New York, 1967.

I. Glicksberg

1. *The abstract F. and M. Riesz theorem*, J. Functional Anal. **1** (1967), 109–122.

G. M. Goluzin

1*. *Geometric theory of functions of a complex variable*, GITTL, Moscow-Leningrad, 1952; English transl. of the 2nd ed., Transl. Math. Mono., vol. 26, Amer. Math. Soc., Providence, R.I., 1969.

Hans Grauert and Ingo Lieb

1. *Das Ramirezsche Integral und die Lösung der Gleichung $\bar{\partial}f = \alpha$ im Bereich der beschränkten Formen*, Complex Analysis (Proc. Conf., Houston, Texas, 1969), Rice Univ. Studies **56** (1970), no. 2, 29–50.

Robert C. Gunning and Hugo Rossi

1. *Analytic functions of several complex variables*, Prentice-Hall, Englewood Cliffs, N.J., 1965.

F. Reese Harvey and R. O. Wells, Jr.

1. *Holomorphic approximation and hyperfunction theory on a C^1 totally real submanifold of a complex manifold*, Math. Ann. **197** (1972), 287–318.

Maurice Heins

1. *On the theorem of Szegö-Solomentsev*, Math. Scand. **20** (1967), 281–289.

Kenneth Hoffman

1. *Banach spaces of analytic functions*, Prentice-Hall, Englewood Cliffs, N.J., 1962.

Lars Hörmander

1. *An introduction to complex analysis in several variables*, Van Nostrand, Princeton, N.J., 1966.

L. Hörmander and J. Wermer

1. *Uniform approximation on compact sets in* $\mathbf{C}^n$, Math. Scand. **23** (1968), 5–21.

Mitsuyuki Itano

1. *On the multiplicative products of distributions*, J. Sci. Hiroshima Univ. Ser. A-I Math. **29** (1965), 51–74.

V. K. Ivanov

1. *The operation of multiplication for distributions of slow growth*, Izv. Vyssh. Uchebn. Zaved. Matematika **1972**, no. 3(118), 10–19. (Russian)
2. *Hyperdistributions and the multiplication of Schwartz distributions*, Dokl. Akad. Nauk SSSR **204** (1972), 1045–1048; English transl. in Soviet Math. Dokl. **13** (1972).

3*. *An associative algebra of elementary generalized functions*, Sibirsk. Mat. Zh. **20** (1979), 731–740; English transl. in Siberian Math. J. **20** (1979).

V. K. Ivanov and T. A. Verzhbalovich

1*. *On the multiplication of Schwartz distributions*, Trudy Inst. Mat. i Mekh. Ural. Nauchn. Tsentr Akad. Nauk SSSR Vyp. 13 (1974), 3–10. (Russian)

M. V. Keldysh

1. *Sur la représentation par des séries de polynomes des fonctions d'une variable complexe dans de domaines fermés*, Mat. Sb. **16(58)** (1945), 249–258. (Russian; French summary)

Norberto Kerzman

1. *Hölder and L^p estimates for solutions of $\bar{\partial} u = f$ in strongly pseudoconvex domains*, Comm. Pure Appl. Math. **24** (1971), 301–379.

G. M. Khenkin

1. *Integral representations of functions holomorphic in strictly pseudoconvex domains and some applications*, Mat. Sb. **78(120)** (1969), 611–632; English transl. in Math. USSR Sb. **7** (1969).
2. *Integral representation of functions in strictly pseudoconvex domains and applications to the $\bar{\partial}$-problem*, Mat. Sb. **82(124)** (1970), 300–308; English transl. in Math. USSR Sb. **11** (1970).
3. *A uniform estimate for the solution of the $\bar{\partial}$-problem in a Weil region*, Uspekhi Mat. Nauk **26** (1971), no. 3(159), 211–212. (Russian)

4*. *Solutions with estimates of the H. Lewy and Poincaré-Lelong equations*, Dokl. Akad. Nauk SSSR **224** (1975), 771–774; English transl. in Soviet Math. Dokl. **16** (1975).

5*. *The Lewy equation and analysis on pseudoconvex manifolds*. I, Uspekhi Mat. Nauk **32** (1977), no. 3(195), 57–118; English transl. in Russian Math. Surveys **32** (1977).

6*. *The Lewy equation and analysis on pseudoconvex manifolds*. II, Mat. Sb. **102(144)** (1977), 71–108; English transl. in Math. USSR Sb. **31** (1977).

J. J. Kohn

1. *Regularity at the boundary of the $\bar{\partial}$-Neumann problem*, Proc. Nat. Acad. Sci. U.S.A. **49** (1963), 206–213.

J. J. Kohn and Hugo Rossi

1. *On the extension of holomorphic functions from the boundary of a complex manifold*, Ann. of Math. (2) **81** (1965), 451–472.

Heinz König and G. L. Seever

1. *The abstract F and M. Riesz theorem*, Duke Math. J. **36** (1969), 791–797.

Walter Koppelman

1. *The Cauchy integral for differential forms*, Bull. Amer. Math. Soc. **73** (1967), 554–556.
2. *The Cauchy integral for functions of several complex variables*, Bull. Amer. Math. Soc. **73** (1967), 373–377.

Adam Korányi

1*. *The Poisson integral for generalized half-planes and bounded symmetric domains*, Ann. of Math. (2) **82** (1965), 332–350.

A. M. Kytmanov

1*. *A criterion for the holomorphy of an integral of Martinelli-Bochner type*, Combinatorial and Asymptotic Analysis, Krasnoyarsk. Gos. Univ., Krasnoyarsk, 1975, pp. 169–177. (Russian)

2*. *On certain properties of exterior differential forms of class B*, On Holomorphic Functions of Several Complex Variables, Inst. Fiz. Sibirsk. Otdel. Akad. Nauk SSSR, Krasnoyarsk, 1976, pp. 184–189. (Russian)

3. *A certain characteristic property of $\bar{\partial}$-closed exterior differential forms*, Uspekhi Mat. Nauk **31** (1976), no. 2(188), 217–218. (Russian)

4*. *On an integral characteristic property of $\bar{\partial}$-closed complex differential forms*, Sibirsk. Mat. Zh. **19** (1978), 788–792; English transl. in Siberian Math. J. **19** (1978).

5. *The representation and product of distributions of several variables by means of harmonic functions*, Izv. Vyssh. Uchebn. Zaved. Matematika **1978**, no. 1(188), 36–42; English transl. in Soviet Math. (Iz. VUZ) **22** (1978).

6*. *A class of multidimensional distributions*, Izv. Vyssh. Uchebn. Zaved. Matematika **1980**, no. 10(221), 23–28; English transl. in Soviet Math. (Iz. VUZ) **24** (1980).

7*. *Some differential criteria for holomorphy of functions in* $\mathbf{C}^n$, Some Questions of Multidimensional Complex Analysis, Inst. Fiz. Sibirsk. Otdel. Akad. Nauk SSSR, Krasnoyarsk, 1980, pp. 51–64. (Russian)

8*. *On the exact calculation of an integral of Martinelli-Bochner type in a ball in* $\mathbf{C}^n$, Uspekhi Mat. Nauk **36**(1981), no. 3(219), 217–218; English transl. in Russian Math. Surveys **36** (1981).

A. M. Kytmanov and L. A. Aĭzenberg

1*. *The holomorphy of continuous functions that are representable by the Martinelli-Bochner integral*, Preprint No. 5M, Inst. Fiz. Sibirsk. Otdel. Akad. Nauk SSSR, Krasnoyarsk, 1975. (Russian) R Zh Mat. **1975** #8Б166.

2*. *The holomorphy of continuous functions that are representable by the Martinelli-Bochner integral*, Izv. Akad. Nauk Armyan. SSR Ser. Mat. **13** (1978), 158–169. (Russian)

A. M. Kytmanov and A. A. Preobrazhenskiĭ

1. *On a problem about the density of the polynomials of a definite form in the space of continuous functions on the boundary of the domain in* $\mathbf{C}^n$, Izv. Vyssh. Uchebn. Zaved. Matematika **1976**, no. 7(170), 50–51; English transl. in Soviet Math. (Iz. VUZ) **20** (1976).

Jean Leray

1. *Le calcul différentiel et intégral sur une variété analytique complexe* (*Problème de Cauchy*, III), Bull. Soc. Math. France **87** (1959), 81–180.

Ingo Lieb

1. *Die Cauchy-Riemannschen Differentialgleichungen auf streng pseudokonvexen Begieten. Beschränkte Lösungen*, Math. Ann. **190** (1970/71), 6–44.

Lu Qi-keng and Zhong Tong-de

1. *An extension of Privalof's theorem*, Acta Math. Sinica **7** (1957), 144–165. (Chinese; English summary)

L. Ja. Makarova, G. Kudaĭberganov and V. P. Cherkashin

1. *A remark on polynomial and linear convexity*, Izv. Akad. Nauk UzSSR Ser. Fiz.-Mat. Nauk **17** (1973), no. 1, 36–41. (Russian)

Bernard Malgrange

1. *Ideals of differentiable functions*, Tata Inst. Fund. Res., Bombay, and Oxford Univ. Press, London, 1967.

2. *Existence et approximation des solutions des équations aux dérivées partielles et des équations de convolution*, Ann. Inst. Fourier (Grenoble) **6** (1955/56), 271–355.

A. Martineau

1. *Sur la topologie des espaces de fonctions holomorphes*, Math. Ann. **163** (1966), 62–88.

2. *Distributions et valeurs au bord des fonctions holomorphes*, Theory of Distributions (Proc. Internat. Summer Inst., Lisbon, 1964), Inst. Gulbenkian Ci., Lisbon, 1964, pp. 193–326.

E. Martinelli

1. *Sopra un teorema di F. Severi nella teoria delle funzioni analitiche di più variabili complesse*, Rend. Mat. e Appl. (5) **20** (1961), 81–96.

S. N. Mergelyan

1. *Uniform approximations to functions of a complex variable*, Uspekhi Mat. Nauk **7** (1952), no. 2(48), 31–122; English transl. in Amer. Math. Soc. Transl. (1) **3** (1962).

N. I. Muskhelishvili

1. *Singular integral equations*, 2nd ed., Fizmatgiz, Moscow, 1962; German transl., Akademie-Verlag, Berlin, 1965.

Nils Øvrelid

1. *Integral representation formulas and* L^p*-estimates for the* $\bar\partial$*-equation*, Math. Scand. **29** (1971), 137–160.

V. P. Palamodov

1. *Linear differential operators with constant coefficients*, "Nauka", Moscow, 1967; English transl., Springer-Verlag, 1970.

P. L. Polyakov

1. *The Cauchy-Weil formula for differential forms*, Mat. Sb. **85(127)** (1971), 388–402; English transl. in Math. USSR Sb. **14** (1971).
2. *Banach cohomology of piecewise strictly pseudoconvex domains*, Mat. Sb. **88(130)** (1972), 238–255; English transl. in Math. USSR Sb. **17** (1972).

I. I. Privalov

1. *Boundary properties of analytic functions*, 2nd ed., GITTL, Moscow, 1950; German transl., VEB Deutscher Verlag Wiss., Berlin, 1956.

John Rainwater

1. *A note on the preceding paper*, Duke Math. J. **36** (1969), 799–800.

Enrique Ramírez de Arellano

1. *Ein Divisionsproblem und Randintegraldarstellungen in der komplexen Analysis*, Math. Ann. **184** (1969/70), 172–187.

Georges de Rham

1. *Variétés différentiables. Formes, courants, formes harmoniques*, Actualités Sci. Indust., No. 1222, Hermann, Paris, 1955.

G. B. Rizza

1. *Su diverse estensioni dell'invariante di E. E. Levi nella teoria delle funzioni di più variabili complesse*, Ann. Mat. Pura Appl. (4) **44** (1957), 73–89.

Ya. A. Roĭtberg

1*. *On boundary values of generalized solutions of elliptic equations*, Mat. Sb. **86(128)** (1971), 248–267; English transl. in Math. USSR Sb. **15** (1971).

A. V. Romanov

1*. *Spectral analysis of the Martinelli-Bochner operator for the ball in* $\mathbf{C}^n$, *and its applications*, Funktsional. Anal. i Prilozhen. **12** (1978), no. 3, 86–87; English transl. in Functional Anal. Appl. **12** (1978).

2*. *Convergence of iterates of the Bochner-Martinelli operator, and the Cauchy-Riemann equation*, Dokl. Akad. Nauk SSSR **242** (1978), 780–783; English transl. in Soviet Math. Dokl. **19** (1978).

A. V. Romanov and G. M. Kenkin

1. *Exact Hölder estimates for the solutions of the* $\bar{\partial}$*-equation*, Izv. Akad. Nauk SSSR Ser. Mat. **35** (1971), 1171–1183; English transl. in Math. USSR Izv. **5** (1971).

Walter Rudin

1*. *Pluriharmonic functions in balls*, Proc. Amer. Math. Soc. **62** (1977), 44–46.

B. V. Shabat

1. *Introduction to complex analysis*, "Nauka", Moscow, 1969. (Russian)

Pierre Schapira

1. *Theorie des hyperfonctions*, Lecture Notes in Math., vol. 126, Springer-Verlag, 1970.

A. I. Serbin

1. *On functions representable by a Martinelli-Bochner integral*, Dokl. Akad. Nauk SSSR **196** (1971), 1276–1279; English transl. in Soviet Math. Dokl. **12** (1971).

G. E. Shilov

1. *Mathematical analysis: Second special course*, "Nauka", Moscow, 1965; English transl., *Generalized functions and partial differential equations*, Gordon and Breach, New York, 1968.

Henri Skoda

1*. *Zéros des fonctions de la classe de Nevanlinna dans les ouverts strictement pseudoconvexes*, C. R. Acad. Sci. Paris Sér. A-B **280** (1975), A1677–A1680.

2*. *Valeurs ou bord pour les solutions de l'opérateur d'', et caractérisation des zéros des fonctions de la classe de Nevanlinna*, Bull. Soc. Math. France **104** (1976), 225–299.

S. L. Sobolev

1*. *Introduction to the theory of cubature formulas*, "Nauka", Moscow, 1974. (Russian)

Elias M. Stein

1. *Boundary behavior of holomorphic functions of several complex variables*, Math. Notes, No. 11, Princeton Univ. Press, Princeton, N.J., and Univ. of Tokyo Press, Tokyo, 1972.

2*. *Singular integrals and differentiability properties of functions*, Princeton Univ. Press, Princeton, N.J., 1970.

Edgar Lee Stout

1*. *The boundary values of holomorphic functions of several complex variables*, Duke Math. J. **44** (1977), 105–108.

Heinz Günther Tillmann

1*. *Darstellung der Schwartzachen Distributionen durch analytische Funktionen*, Math. Z. **77** (1961), 106–124.

A. G. Vitushkin

1. *Necessary and sufficient conditions a set should satisfy in order that any function continuous on it can be approximated uniformly by analytic or rational functions*, Dokl. Akad. Nauk SSSR **128** (1959), 17–20. (Russian)

V. S. Vladimirov

1. *Methods of the theory of functions of many complex variables*, "Nauka", Moscow, 1964; English transl., M.I.T. Press, Cambridge, Mass., 1966.

2. *The Laplace transform of generalized functions of slow growth*, Itogi Nauki: Sovremennye Problemy Mat., Vol. 1, VINITI, Moscow, 1973, pp. 61–84; English transl. in J. Soviet Math. **3** (1975), no. 2.

3. *On the construction of envelopes of holomorphy for regions of a special type, and their applications*, Trudy Mat. Inst. Steklov. **60** (1961), 101–144; English transl. in Amer. Math. Soc. Transl. (2) **48** (1965).

4. *Problems of linear junction of holomorphic functions of several complex variables*, Izv. Akad. Nauk SSSR Ser. Mat. **29** (1965), 807–835; English transl. in Amer. Math. Soc. Transl. (2) **71** (1968).

5*. *Equations of mathematical physics*, "Nauka", Moscow, 1967; English transl., Marcel Dekker, New York, 1971.

Dietmar Vogt

1. *Distributionen auf dem* $\mathbf{R}^n$ *als Randverteilungen holomorpher Funktionen*, J. Reine Angew. Math. **261** (1973), 134–145.

E. A. Volkov

1*. *Boundaries of subdomains, Hölder weight classes and solutions in these classes of the Poisson equation*, Trudy Mat. Inst. Steklov. **117** (1972), 75–99; English transl. in Proc. Steklov Inst. Math. **117** (1972).

Barnet M. Weinstock

1. *Continuous boundary values of analytic functions of several complex variables*, Proc. Amer. Math. Soc. **21** (1969), 463–466.
2. *An approximation theorem for $\bar{\partial}$-closed forms of type* $(n, n-1)$, Proc. Amer. Math. Soc. **26** (1970), 625–628.

3*. *Uniform approximation by solutions of elliptic equations*, Proc. Amer. Math. Soc. **41** (1973), 513–517.

John Wermer

1. *An example concerning polynomial convexity*, Math. Ann. **139** (1959), 147–150; addendum, ibid. **140** (1960), 322–323.

A. P. Yuzhakov and V. P. Krivokolesko

1. *Certain properties of linearly convex domains with smooth boundaries in* $\mathbf{C}^n$, Sibirsk. Mat. Zh. **12** (1971), 452–458; English transl. in Siberian Math. J. **12** (1971).

SUBJECT INDEX

INDEX OF SYMBOLS